DESIGN and USE of PRESSURE SEWER SYSTEMS

David Thrasher, PE
Thrasher Engineering
Rogers, Arkansas

LEWIS PUBLISHERS

Library of Congress Cataloging-in-Publication Data

Thrasher, David.
  Design and use of pressure sewer systems.

  Bibliography: p. 105.
  Includes index.
  1. Sewerage—Fluid dynamics. I. Title.
II. Title: Pressure sewer systems.

TD658.T48   1987   628.3   86-27340
ISBN 0-87371-070-3

FLOOD DAMAGE

2 0 JAN 1996

COPYRIGHT © 1987 by LEWIS PUBLISHERS, INC.
ALL RIGHTS RESERVED

Neither this book nor any part may be reproduced or transmitted in any form or by any means, electronic or mechanical, including photocopying, microfilming, and recording, or by any information storage and retrieval system, without permission in writing from the publisher.

LEWIS PUBLISHERS, INC.
121 South Main Street, P.O. Drawer 519, Chelsea, Michigan 48118

PRINTED IN THE UNITED STATES OF AMERICA

Preface

The purpose of this book is to provide the reader with an overview of the design and use of pressure sewer systems. It attempts to consolidate the existing literature, along with the author's experience, into a single useful reference.

There is an expanding audience that wants more information about pressure sewers, their design and use, advantages and disadvantages, operation and maintenance, front-end cost savings, and the possibilities created by pressure sewers for making sewers available in areas where on-site disposal has proven or may prove to be undesirable.

This reference can be used as a technical design guide by engineers, or as an information source for others interested in the use of pressure sewer systems. It should prove useful to design or consulting engineers, municipalities, sewer districts, sanitary engineers, developers, and government and regulatory agencies, as well as pump and equipment manufacturers.

David Thrasher
Rogers, Arkansas

Acknowledgments

The reader will note that I have quoted Mr. Bill Bowne of Eugene, Oregon extensively; that is because I consider Mr. Bowne to be the foremost expert on pressure sewer technology. I would like to take this opportunity to thank Mr. Bowne for providing much of the guidance that has led to my experience in this field.

My thanks are also extended to Mr. James F. Gore of Bella Vista Village, Arkansas. Mr. Gore's insight and guidance through the years have been invaluable to my overall experience. Many thanks also go to Mr. Johnny Overton, Mr. James Kreissl, Mr. Robert Langford, and Mr. Newton White, all of whom are leading experts on pressure sewer systems.

 **David Thrasher** attended the University of Arkansas, where he obtained a Bachelor of Science degree in civil engineering in 1973. After two years with the Tennessee Valley Authority as a field engineer, he returned to the University of Arkansas for graduate studies in civil engineering with an emphasis on environmental engineering, and obtained his Master of Science degree.

In 1976, Mr. Thrasher was employed by Cooper Communities, Inc., a large community developer based in Arkansas. During his employment there, Mr. Thrasher was instrumental in establishing design criteria for pressure sewer systems in Hot Springs Valley, Arkansas; Bella Vista Village, Arkansas; and Tellico Village, Tennessee. He also served as a consultant helping to establish design criteria for a pressure sewer system in Palm Coast, Florida.

Mr. Thrasher has conducted numerous seminars about pressure sewer systems, some of which have been sponsored by the University of Arkansas, the Arkansas Department of Health, the Arkansas Department of Pollution Control and Ecology, and the Environmental Protection Agency. He currently has his own consulting engineering firm in Rogers, Arkansas.

Contents

1. Introduction 1
2. History of Pressure Sewers 5
3. General Information Concerning Pressure Sewer Systems 11
4. Preliminary Design Concepts and Considerations . 27
5. Final Design Considerations 39
6. Design Methodology 61
7. Equipment and Material Considerations 77
8. Characteristics of Pressure Sewage 91
9. Operation and Maintenance 97

Bibliography 105

Glossary 115

Index 121

List of Figures

1.1	Cost of Gravity Sewers as a Function of Population Density	4
3.1	Typical Grinder Pump Installation	12
3.2	Typical Septic Tank Effluent Pump (STEP) Installation	13
4.1	Anticipated Frequency of Pumping Interceptor Tanks	33
4.2	Comparison of Pump Performance Curves for Semi-Positive Displacement Pumps and Centrifugal Pumps	38
5.1	Suggested Design Flows for 150 GPD Average Daily Flow	43
5.2	STEP Pressure Sewer Design Chart	44
5.3	Recommended Design Flow from "Design and Specification Guidelines for Low Pressure Sewer Systems" (13)	45
6.1	Plan View of Gravity Sewer Design for Armstrong Subdivision	62
6.2	Profile of Gravity Sewer Design for Armstrong Subdivision	63
6.3	Plan View of Pressure Sewer Design for Armstrong Subdivision	65
6.4	Profile of Pressure Sewer Design for Armstrong Subdivision	66
6.5	Hydraulic Gradient Calculation and Pump Selection	68

List of Tables

1.1	General Disadvantages of Conventional Gravity Wastewater Collection Systems	2
3.1	General Advantages of Pressure Sewer Systems over Conventional Gravity Sewer Systems	14
4.1	A Comparison of the Relative Costs of Grinder Pump and STEP Pressure Sewer Systems	28
4.2	Duration of Power Outages as Reported by the Federal Power Commission	35
5.1	Maximum Number of Grinder Pump Cores Operating Simultaneously	41
6.1	Gravity Sewer Construction Cost Estimate for Armstrong Subdivision	73
6.2	Pressure Sewer Construction Cost Estimate for Armstrong Subdivision	74
6.3	Economic Present Worth Analysis of Gravity Sewer Cost for Armstrong Subdivision	75
6.4	Economic Present Worth Analysis of Pressure Sewer Cost for Armstrong Subdivision Assuming All Capital Cost for First Year	75
6.5	Economic Present Worth Analysis of Pressure Sewer Cost for Armstrong Subdivision Assuming Delayed Housing Growth	75
8.1	Treatment Facility Influent Characteristics	91
8.2	Selected Characteristics of Pressure Sewers	92
8.3	Average Septic Tank Influent and Effluent Characteristics	92

8.4	Wastewater Characteristics of the Glide, Oregon Pressure Sewer System	93
8.5	Selected Wastewater Characteristics for STEP and Grinder Pump Systems	94
8.6	Albany, New York Grinder Pump Wastewater Characteristics	94
8.7	Household Wastewater Characteristics	95

1
Introduction

Currently, the field of wastewater engineering is in an extremely dynamic period of development. Due to the high cost of wastewater collection systems, engineers are reevaluating old ideas concerning wastewater collection and new concepts are being formulated.

The traditional mode of collection for wastewater has been gravity sewers. One problem with this traditional solution has been the obvious fact that gravity sewers must slope downhill. This creates a situation where deep cuts are often required and large, expensive pump stations are often necessary. In addition, the design engineer is faced with a situation where the location of any new collection main is very limited. Many times, these disadvantages will mean that areas with extreme variations in terrain or with other limitations will remain unsewered, sometimes to the detriment of public health. All of these disadvantages serve to increase the overall cost of any wastewater collection system.

Table 1-1 provides the reader with a general list of disadvantages of conventional gravity wastewater collection systems.

For most wastewater disposal projects, the collection system is usually by far the most expensive portion of the project, while the treatment plant is usually a comparatively small expense. In many rural systems, it is common

2   PRESSURE SEWER SYSTEMS

Table 1-1. General Disadvantages of Conventional Gravity Wastewater Collection Systems

1) Infiltration and inflow are commonly encountered.
2) Downhill slope must be maintained at all times.
3) Large, expensive pump stations are required in those instances where downhill slopes cannot be maintained.
4) Large diameter (8″ or larger) pipe is required to transport the solids commonly found in wastewater.
5) There is very little flexibility in the location of gravity sewer collection mains.
6) Deep utility trenches are often required which leads to such expensive construction techniques as trench shoring and dewatering.

for 80 percent or more of the total cost of a sewer system to lie in the collection system which means that less than 20 percent of the cost would be in the treatment facilities.[4,10] Areas which are more urban in nature, where homes are very close together and where topographic obstacles are not severe, are usually amenable to conventional sewerage systems. However, in sparsely settled communities, in hilly terrain, or in areas where rock excavation or groundwater present problems, costs represented by the collection system can become too great to bear.

Since most on-site wastewater disposal systems cost significantly less than centralized sewer, the on-site approach has been generally employed in areas where on-site disposal is applicable. Difficulties have arisen in areas where conventional on-site systems have failed due to unfavorable soil conditions. The typical result from this condition has almost invariably been a recommendation to provide central sewer for the community. Implementation of this recommendation has been dependent on the financial status of the community, availability of federal grants, and public attitude. Experience has shown that the cost of conventional sewers is extremely high for most small com-

INTRODUCTION 3

munities. In fact, it is not uncommon to see construction cost estimates in excess of $10,000 per home.[4]

The overall cost of any wastewater collection system project is important; however, the cost per user is even more important. Without addressing existing discussions on the merits and demerits of federal grant programs and centralized collection and treatment systems, it is sufficient to note that the cost of conventional sewers is extremely high for most small communities. Also, since the cost of the conventional collection system generally represents a majority of the total system capital cost in rural areas, cost reductions in the collection system can have a significant impact on the overall cost of a sewage system. Figure 1-1 illustrates the inverse relationship between cost and population density, which is primarily explained by the greater length of sewer main per contributor. Other explanations for the high capital cost of conventional wastewater collection systems are regulations which limit the smallest sewer pipe diameter and grade and alignment requirements which often result in deep cuts and expensive lift stations.

Due to the high cost of conventional gravity sewers, many areas remain unsewered today, and these areas have generally retained septic tank-soil absorption systems as the means of wastewater disposal. Even though septic tank-soil absorption systems have an enviable record of performance, many limitations exist. Soil type and depth must be adequate, and population density should remain low. It has been estimated that as much as 65% of the United States contains sufficient limitations on the use of septic tank-soil absorption systems that could severely limit their effectiveness.[9]

For an area that is not conducive to the on-site disposal of human waste and cannot afford the high cost of conventional wastewater collection systems, innovative collection alternatives should be considered. One such alternative to consider should be a pressurized sewer system.

4 PRESSURE SEWER SYSTEMS

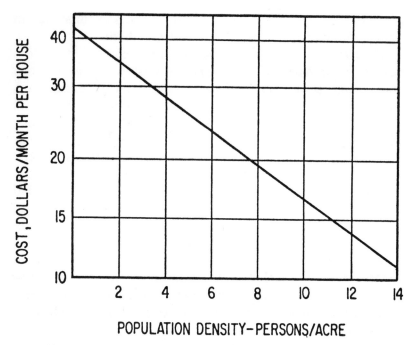

Figure 1-1. Cost of gravity sewers as a function of population density.

2
History of Pressure Sewers

The concept of grinder pressure sewers was first proposed by Dr. Gordon M. Fair, Professor of Sanitary Engineering at Harvard University, in 1954. His idea of pressure sewers was part of a combined sewer system that would entail hanging a pressurized sewer pipe within a gravity storm sewer pipe. The object was to pump the domestic sewage through the pressure pipe and convey storm water by gravity, thus eliminating the need to build a separate sanitary sewer system in some highly populated areas.

The first published experience in the design and installation of pressure sewers came from an engineer in Kentucky who designed a pressure sewer system for installation in a section of Radcliffe, Kentucky. This small demonstration project, installed in 1964, served 48 homes.[3] This system was later dismantled and replaced by a gravity sewer system.[2]

In 1966, the American Society of Civil Engineers (ASCE) initiated a study under a federal grant in an attempt to prove Dr. Fair's theory concerning pressure sewers. Three years later, ASCE published a final report which concluded that Dr. Fair's concept of combined sewers was not economical. However, the report did state that

6 PRESSURE SEWER SYSTEMS

small-diameter, low-pressure sewer systems offered a potential cost-effective alternative to conventional gravity sewers, especially where installation of gravity sewers was considered uneconomical or infeasible.[43]

The oldest known pressure sewer system in existence today was installed in 1968 and serves a permanent houseboat community located on the Columbia River near Portland, Oregon. In this system, each houseboat has a 1/2-horsepower sewage ejector (solids handling) pump located in a basin attached to the side of the houseboat. There are about 700 pumps in operation, with 150 pumps operating in parallel within one section of the system. This system has operated successfully for over 16 years, and it is one example of the performance of a pressure sewer system utilizing centrifugal pumps in parallel operation.

The subsequent development of pressure sewer systems has fallen into two categories; first, grinder pump pressure sewer systems, and second, pumped effluent or Septic Tank Effluent Pumping (STEP) pressure sewer systems.

2.1 GRINDER PUMP PRESSURE SEWER SYSTEMS

As part of the ASCE study, General Electric Company developed a grinder pump in 1967–68 that was to be used as part of the field demonstration. Due to budget considerations, this particular field demonstration was never accomplished. However, the New York Department of Environmental Conservation's Research & Development Unit decided to sponsor a demonstration with United States Environmental Protection Agency (USEPA) grant funding to determine the feasibility of a grinder pressure sewer system at Albany, New York.[5]

In addition to Albany, New York, the USEPA also funded full scale evaluations of grinder pump pressure sewer systems at Phoenixville, Pennsylvania,[6] Grandview Lake, Indiana,[7] and Bend, Oregon.[67]

The Albany, New York Pressure Sewer System consisted of 12 grinder pumps. The pressure main for that system was oversized to allow for all units to operate simultaneously. It was reported that subsequent accumulations of grease and fibrous materials within the pipes reduced some pipe cross-sectional areas by as much as 40%. In addition, the wastewater was found to be more concentrated than normal, apparently due to lack of infiltration in the pressure sewer system.[5] Indirect evidence of reduction of the pipe cross-sectional area by grease was also found at Phoenixville.[6] Grandview Lake also reported grease problems, faulty operation of automatic air release valves, and grease buildup on flow measuring devices at the wastewater treatment plant.[4]

The Grandview Lake Project was installed near Columbus, Indiana in late 1971. The system was a combination grinder and effluent pressure sewer demonstration project, and it consisted of 58 house connections in the first phase. The majority of the pumps were grinders along with a few STEP pumps.[7]

A more recent example of grinder pump system installation is within the Pennsylvania Pocono Mountains. Here there are numerous private land developments which consist primarily of seasonal occupancy dwellings. As of 1982, over 750 units had been installed and numerous future units were being planned. Pressure sewer was chosen for this locale because it provided advantages in land planning and because of economic advantages which included lower initial capital cost. Pressure sewers were well suited to the rocky and hilly terrain and to the layout of houses in interspersed clusters at different locations within the mountains. Early development in this area was served by gravity sewers, but after examination of the advantages of pressure sewers, most development since 1974 has been grinder or effluent pressure sewer.[2]

8 PRESSURE SEWER SYSTEMS

2.2 STEP PRESSURE SEWER SYSTEMS

Pumps to convey septic tank effluent have been in use for many years in individual home applications. In these situations, the unit would normally pump the effluent from a septic tank to a soil absorption field some distance from the home and normally at a higher elevation. Cooper Communities, Inc., a large community developer in Arkansas, reports that they have been installing systems such as this within their developments since 1958.[66]

Rose[8] is generally credited as one of the first proponents of STEP systems. He proposed this concept to the Farmers Home Administration as a solution to sewage problems in rural communities as early as 1967.

At about the same time, Harold Schmidt, of the General Development Corporation Utility Company, Miami, Florida, initiated a feasibility study of pumping septic tank effluent as a basic premise for a pressure sewer system. Based on this investigation, a small pressure system was installed at the Port Charlotte, Florida development in 1970. Forty homes were connected to the system and the pressure sewer from these homes discharged directly into an extended aeration treatment plant. This was the first treatment plant to be designed specifically for treating septic tank effluent. A year later, General Development Corporation installed another effluent pressure sewer system at their Port St. Lucie, Florida project. This pressure sewer system consisted of 75 homes discharging into a municipal wastewater system with an activated sludge sewage treatment facility. This pressure sewer system has since been expanded to approximately 300 homes.

One of the first field evaluations to determine the feasibility of pumping septic tank effluent through small pipelines as part of a pressure sewer system was conducted as part of the Grandview Lake project. Twenty-seven submersible centrifugal effluent pumps were installed at Grandview Lake at individual homes around the lake.

HISTORY OF PRESSURE SEWERS

These pumps were installed in steel basins outside the septic tank. The remainder of the pumps (57 units) were semi-positive displacement grinder pumps.[7] Based on the performance of grinder and effluent pumps in this demonstration pressure sewer system, the design engineer subsequently recommended that septic tank effluent pumping be used in future pressure sewer installations.[71]

Kenneth Durtschi, a consulting engineer in Coeur d'Alene, Idaho, designed a successful STEP system for the Preist Lake, Idaho area in 1972. This system includes 550 homes and two treatment plants and is served satisfactorily by one service man.[2,9,12]

Also in the early seventies, William Bowne, an engineer with Douglas County, Oregon, conducted an extensive study to determine the best method for sewage disposal in the Glide-Idleyld Park Communities.[10] After visiting several pressure sewer system projects throughout the county, Mr. Bowne successfully convinced Douglas County Officials to design and install an effluent pressure sewer system for those communities. A detailed report was published in 1975,[11] and Mr. Bowne has subsequently been acknowledged as one of the foremost experts on pressure sewer technology.

Another significant STEP system is located at Manila, California in the northwest portion of the state. The system was designed by the consulting engineering firm of Winzler and Kelly, with consultation from William Bowne.[12] Funding for the system was obtained through the California State Water Resources Control Board to construct a pilot alternative system and collect much needed data on system operation and maintenance requirements and costs. The STEP system was selected because it was judged to be the most easily constructed in local unstable soil conditions and the most economical system available, and because of reduced treatment costs and the advantage of additional holding capacity within the septic or interceptor tanks. The conditions in Manila were

10 PRESSURE SEWER SYSTEMS

high groundwater, high population density with closely spaced homes, and failing soil absorption systems. A comprehensive report was prepared on this system by the design engineering firm in conjunction with the State of California Water Resources Control Board.[12]

2.3   EXISTING PRESSURE SEWER SYSTEMS

Rezek, Henry, Meisenheimer, and Gende, Inc. has prepared a report for the USEPA entitled "Investigations of Existing Pressure Sewer Systems."[79] In that report, they note 78 existing pressure sewer systems located in 26 states and in the country of Canada. A brief review of this reference will show that pressure sewer systems are now widely used. For this reason, pressure sewer technology is now widely used. For this reason, pressure sewer technology is now considered as a standard means of wastewater collection, and it should no longer be considered experimental technology.

3
General Information Concerning Pressure Sewer Systems

In the research for new sewage collection and transmission systems, pressure sewer systems have been widely recognized as one alternative worth further study. Pressure sewer systems can be utilized to serve difficult areas, and pressure sewer mains have successfully been built at a fraction of the cost of conventional sewers.

The two major types of pressure sewer systems are the grinder pump system and the septic tank effluent pumping (STEP) system. These different types of systems are depicted in Figures 3-1 and 3-2. From these figures it is obvious that the major differences between these systems are in the on-site equipment, but some differences may also exist in the pressure mains and appurtenances, mainly due to the solids and grease that must be transported in a grinder pump system. As one can see after examining Figures 3-1 and 3-2, the grinder pump pressure sewer system is based upon the premise that the grinder pump macerates and pumps all liquids and solids, including grease. The STEP system, on the other hand, retains much of the solids and grease within the individual septic or interceptor tanks. It is also significant that neither type of pressure sewer system requires any modification of the plumb-

12 PRESSURE SEWER SYSTEMS

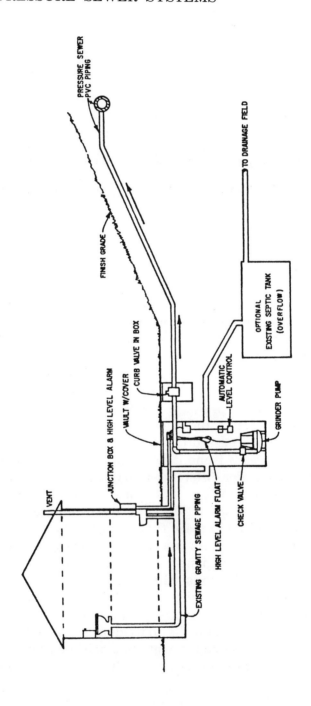

Figure 3-1. Typical grinder pump installation.

GENERAL INFORMATION 13

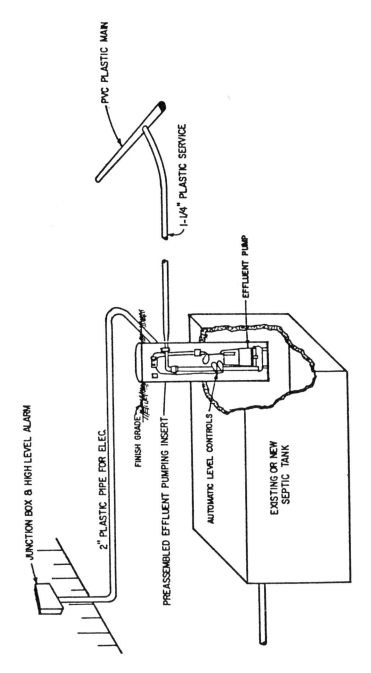

Figure 3-2. Typical septic tank effluent pump (STEP) installation.

14   PRESSURE SEWER SYSTEMS

Table 3-1. General Advantages of Pressure Sewer Systems Over Conventional Gravity Sewer Systems

1. A major portion of the investment for the sewer system is the on-lot pressurization system, and this investment will not be required until a house is constructed on any given lot.
2. Unlike gravity sewers, pressure sewers do not have severe restrictions concerning minimum pipe size. Minimum pipe diameters in pressure sewer systems are commonly 1¼" to 1½" which compares to a generally accepted figure of 8" minimum pipe size for gravity sewers.
3. Pressure sewers do not have the downhill slope requirements of gravity sewers, and this serves to eliminate deep trenches.
4. Infiltration is practically eliminated.
5. Pressure sewers have a great deal of flexibility in design and construction.

ing within individual households. This allows use of pressure sewer systems without changing the general routine and habits of the users.

3.1   ADVANTAGES OF PRESSURE SEWER SYSTEMS

Numerous advantages of pressure sewers over conventional gravity sewers have been reported. General advantages are shown in Table 3-1.

3.1.1   Economic Advantages

In many instances, it can be shown that pressure sewer systems are less expensive than conventional gravity sewer systems. An economic comparison of these types of systems should only be made when taking all costs into consideration, including operation and maintenance costs.

When such a comparison shows conclusively that a pressure sewer system is economically superior, then such a system should be seriously considered.

One economic fact is generally always true with pressure sewer systems, and that is the cost of the individual pressurization units represents a major portion of the investment in the entire pressure sewer system. This is true whether grinder pump systems or STEP systems are chosen. This can become an extremely important advantage economically due to the fact that the pressurization unit is not required until sewer service must be provided. This means that a major portion of the investment for the pressure sewer system is not required until houses are constructed. Therefore, expenditures for portions of a pressure sewer system can possibly be delayed into the future, and this is extremely important within retirement and recreational communities, where it may be years before a house is constructed on any given lot, and some lots will never have houses built upon them.

Langford[9] has recognized that one of the most important economic advantages of pressure sewer systems is the low front end investment, especially in developments that will have a slow buildup rate of homes. New recreational lake developments have an average annual home building rate which can be as low as 2%. Consequently, it may take up to fifty years or more before homes are built on all lots.

In the case of a pressure sewer system, the small force mains can be installed in the beginning with a minimum investment. When the lot owner builds his home, he would be required to buy his pumping system and hook up to the forcemain, but this cost would be deferred until construction of the home actually took place. In addition, many lot owners choose to never build upon their lots, and this contributes to the beneficial cost defferal factor for the lot owner and the developer. For relative costs of pressure sewer systems versus conventional gravity sewer systems, see Section 6.2-Relative Costs of Pressure Sewer Systems.

The construction of pressure sewer mains is very similar to the construction of water mains. Due to the fact that the liquid is pressurized, gravity flow is not necessary, and the strict alignment and slope restrictions for gravity sewers can be discarded. Pipes can essentially be laid in any location, and this means that extensions for a pressure sewer system can be made in the street right-of-way at a relatively small cost without damage to existing improvements or to the natural environment. This also means that specialized easements will not be required since the street right-of-way can always be utilized, and saving the cost of easements and their procurement is a major advantage both economically and politically.

The piping for pressure sewer systems consists of closed pressurized conduits which will virtually eliminate all infiltration into the system. The elimination of infiltration can reduce the initial capital costs of a sewer system since pipe sizes, pump stations, and treatment plants do not have to be hydraulically sized to handle this additional load. This allows the engineer to design smaller pipes, pump stations, and treatment plants, and this can be a tremendous economic advantage. In addition, the elimination of infiltration reduces system operation and maintenance costs and treatment plant operation and maintenance costs. A smaller amount of liquid flowing through the system means that pumps will operate less which saves electricity costs and adds to the effective life of the pumps. In addition, smaller liquid flows through the treatment plant will also serve to lengthen the effective life of the hydraulic appurtenances within the treatment plant. Such increases in effective life of equipment will serve to extend operation and maintenance costs over longer periods of time and will also serve to delay replacement costs further into the future. The extension and delay of such costs produce true economic advantages to the user.

In general, it suffices to say that engineering costs for pressure sewer systems expressed as a percentage of total construction costs are higher than engineering costs for conventional gravity sewer systems. The extensive hydraulic calculations required for pressure sewer systems combined with the quality control and inspection required, especially for the individual pressurization units, both contribute to this generally higher percentage. However, there have been instances where the construction costs for pressure sewer systems were so much lower than a comparable gravity sewer system that the engineering cost of the pressure sewer system was also lower inspite of being a higher percentage of construction cost. In any case, the overall engineering cost is one factor that should be considered in evaluating whether or not to select a pressure sewer system, and there have been cases where the overall engineering costs for a pressure sewer system have been less than for a comparable gravity sewer system.

Another large economic advantage to the authority operating the pressure sewer system is the fact the electric bills for the individual pressurization units are paid by the homeowner. The pressurization unit is connected to an electric circuit from the house it serves, and the resulting increase to the homeowner's electric bill is minimal. In comparison, this small increase in each homeowner's electric bill often replaces large electric bills for central pump stations, and pressure sewer systems eliminate or reduce the need for such central pump stations. In any case, the operating authority would be required to pay large electric bills for central pump stations, and any reduction in these costs would be a definite advantage.

Pressure sewer systems allow a great degree of flexibility in the location of treatment plants. This advantage can serve to reduce long outfall lines and also helps to avoid locations which would require extensive site work. In addition, it is much easier to centralize the location of the treatment plant, and this helps to reduce pipe sizes in the collection system.

18   PRESSURE SEWER SYSTEMS

3.1.2 Design Advantages

The flexibility in layout of pressure sewer systems creates a definite advantage during the design process. As previously stated, pressure sewer mains can be relocated by the designer in order to miss any obstacle that could create problems during either the design or construction process. In addition, the fact that special easements are rarely required significantly reduces office calculations and field work by survey crews.

3.1.3 Construction Advantages

Small diameter PVC plastic pipe is almost exclusively used in pressure sewer systems, and this type of pipe is much lighter than pipe normally utilized within conventional gravity sewer systems. This lighter pipe may be carried by hand which often frees equipment to continue trenching thus allowing faster pipe installation. PVC pipe also has the capability of specifying longer laying lengths with fewer joints, both of which often serve to decrease construction costs.

Less rock excavation is required in the construction of pressure sewers, since pipes are placed at shallow depths similar to water mains, and since the pipe routing may be easily changed to avoid rock. Less rock excavation means less cost to construct the system and also means that there is less chance of job fatality and injury due to the use of explosives.

Less dewatering is required in utility trenches because pipes are placed at shallow depths and the pipe routing may be changed to avoid high ground water areas. Also, it should be noted that the ability to avoid high ground water areas is a definite advantage for pressure sewer systems in lakeshore oriented developments. When development occurs adjacent to lakes, it is often necessary to place a gravity sewer main along the lakeshore due to the fact that gravity sewers must be placed at lower elevations, especially as they descend through a development. The placement of a gravity sewer next to a lake will subject it to potential infiltration and inflow from the lake. This situation could also subject the lake to a pollution problem if the gravity sewer were to overflow. Conversely, the flexibility of pressure sewers allows them to be located in the road easement, which would normally be some distance away from the lakeshore. Also, aesthetic considerations are much better when sewer main installation does not have to be along the lakeshore. Valuable lakeshore property is not disturbed and the lake does not have to be lowered for construction of the main.

No manholes are required in the construction of pressure sewer systems. Instead, valve assemblies are used and may be connected to piping before being lowered into the trench. Manholes are generally considered to be expensive structures which are often the sources of problems during construction, especially the deeper ones, and the elimination of the need for manholes is a definite advantage for pressure sewers.

Road borings are less time consuming and less expensive for pressure sewer systems. The depth of the trench on each side of the road bore will be less due to the shallower depth of pressure sewer mains. In addition, the bore itself will usually be much smaller because smaller pipe is utilized, and the alignment of the bore itself will not be critical because pipe alignment is not critical.

No shoring is required during the construction of pressure sewers due to the fact that shallow trenches are used, and this creates less chance of construction accidents or fatalities. This can also result in major cost savings in cases where the alternative is an extremely deep gravity sewer since excavation costs tend to increase exponentially with greater depth. Service connections are easier due to the fact that watertight taps can be made with simple tools and that no precise vertical or horizontal alignment is required of service line connections to the pressure sewer main.

3.1.4 Treatment Advantages

STEP pressure sewer systems have a definite advantage at the treatment plant due to the removal of grease and solids in each interceptor tank. It has been estimated that these interceptor tanks retain 50% to 70% of the BOD and Total Suspended Solids from the wastewater generated by each facility served.[53] In addition, examination of data reported by Kreissl[4] and Bowne[77] indicates that interceptor tanks retain approximately 80% of the grease generated by each facility served. The removal of these constituents from the wastewater stream can have definite advantages at the treatment plant. For example, there is no need for grit removal or preliminary settling of solids since settleable solids are removed at the interceptor tanks. In addition, fewer provisions for the removal of organics will be required due to the BOD removed at the interceptor tanks.

Both grinder pump and STEP systems have the overall advantage of much less infiltration than conventional wastewater collection systems. This can be a tremendous advantage at the treatment plant because of the potential reduction in flow through the facility. As an example of this potential reduction in flow, one widely accepted reference recommends that sewer systems be designed on the basis of not less than 100 gallons wastewater flow per capital per day, and the reference states that this figure is assumed to cover normal infiltration.[48] Another accepted reference recommends that pressure sewer systems be designed for 70 gallons of wastewater flow per capita per day.[13] In addition, Bowne[11] has successfully designed pressure sewer systems for 60 gallons of wastewater flow per capita per day, and Cooper Communities, Inc.[27] designed

GENERAL INFORMATION 21

the pressure sewer system at Tellico Village, Tennessee for 50 gallons of wastewater flow per capita per day. A review of these references shows that flow to a treatment facility can be reduced by as much as 30% to 50% by eliminating infiltration. This flow reduction equates to a smaller total hydraulic capacity at the treatment plant which will allow a reduction in the size of all the treatment units contained within the plant which are sized based upon hydraulic flow.

For pressure sewer systems utilizing centrifugal pumps, it is possible to lower the overall system peak flows which can also be an advantage at the treatment plant. Peak flow reduction is accomplished by reducing system pipe sizes to an absolute minimum thereby limiting the amount of flow that can accumulate in the system. Centrifugal pumps make this possible due to their inherent relationship between flow and pressure. As pressures increase, flow decreases, eventually to a point where no flow is produced by the pump. This condition is known as the shutoff head, and it is not desirable to occur constantly; however, it is certainly acceptable to occur occasionally, especially if sufficient storage capacity is available to accommodate the flow into the pressurization unit while the pump is waiting for system pressures to reduce. A more detailed discussion of this phenomenon can be found in Section 4.5-A Comparison of Semi-Positive Displacement Pumps with Centrifugal Pumps.

Any potential reduction in peak flow as described in the preceding paragraph will have a twofold economic advantage. The first and most obvious advantage is the reduction in pipe sizes in the collection system. This will reduce costs because smaller pipes obviously cost less than larger ones. The second advantage of a reduction in peak flow will occur at intermediate pump stations and at the treatment plant. The hydraulic design of pump stations and wastewater treatment plants are based upon peak flows, and any reduction in peak flows will mean a corresponding reduction in the size of these facilities. This reduction in size will mean a reduction in cost.

Thus we have seen that pressure sewer systems can have an effect on costs incurred at the wastewater treatment facility. STEP pressure sewer systems can significantly reduce the biological and solids loading to the wastewater treatment facility which can be utilized as a significant economic advantage. In addition, the lack of infiltration in pressure sewer systems combined with the potential for peak flow reduction in most pressure sewer systems can cause a reduction in treatment plant size due to reduced hydraulic loadings. This can also be a significant economic advantage.

3.1.5 Operation and Maintenance Considerations

In almost every instance, an operating authority should be required for a pressure sewer system. This is in recognition of the fact that pressure sewer systems have many mechanical components, and an operating authority is needed to assure proper operation of all components within the system. The operating authority should have a well conceived operation and maintenance plan which will serve as a guide to the proper operation and maintenance of the system. Some of the aspects of this plan should be as follows:

1) Public education.
2) Inspection of any new installations or connections.
3) Proper monitoring of the system.
4) Schedule of inspections and/or pumping of the on-lot pressurization units.
5) Equipment and labor that will be required to perform all operation and maintenance.
6) An inventory list of spare pumps and replacement parts that should be kept on hand at all times.
7) Emergency plan which will provide contingencies in case of power failures, floods, tornados, etc.

It should be noted that public education should be considered as a top priority of the operating authority. An informed public will assist the authority in assuring proper operation and maintenance. The authority should take every opportunity to make certain the public knows who to call in case of any emergency. In addition, it would be very helpful if the public knew how to react in case of specific types of emergencies. For example, each customer should know to curtail water use in the home if anything is wrong with their pressurization unit. Also, curtailed water use in the case of any type of emergency will provide the authority with more time in which to react to the emergency.

In general, pressure sewers eliminate the need for manholes and for large, conventional sewer pump stations. This is an advantage for the operation and maintenance of the system since inspection of manholes and pump station wetwells is usually considered as an undesirable task and can often be considered dangerous. One other operation and maintenance advantage is the fact the pumps utilized within pressure sewer systems are small and relatively easy to handle and repair. This is in contrast to normal wastewater handling pumps which are usually very large and which require more technical knowledge for maintenance repair purposes.

The disadvantage of pressure sewers in relation to operation and maintenance can be considered to be the number of pumps involved. The pump and controls contained within each pressurization unit within a pressure sewer system are mechanical in nature and are therefore subject to failure. The process of preventing potential failures and the process of repairing failures when they occur is part of the operation and maintenance procedure, and this can create additional responsibilities for operations personnel. However, it has been shown that a well conceived operation and maintenance plan combined with a comprehensive public education program can minimize these additional responsibilities. In addition, when the aforementioned advantages of no manholes, very few large pump

24 PRESSURE SEWER SYSTEMS

stations, and small, easily repairable pumps are considered, the operation and maintenance responsibilities of a pressure sewer system can often be relatively equal to those of a corresponding gravity sewer system.

3.2 DISADVANTAGES OF PRESSURE SEWER SYSTEMS

Operation and maintenance costs for pressure sewer systems are often higher due to the number of pumps involved. Each pump will be subject to eventual maintenance, and this cost should be included when considering pressure sewers. In some cases, operation and maintenance costs may be less if the alternative is one or more large conventional wastewater pumping stations, and this should also be taken into consideration.

One development in Arkansas, which contains both pressure sewers and conventional gravity sewers, has estimated that the operation and maintenance costs for pressure sewers are only about 14 percent higher. This is based on 1985 approximate operation and maintenance costs of $3.31 per month per customer for gravity sewer and $3.77 per month per customer for pressure sewer. These figures are estimates only and should only be used by the reader to gain an overall feeling as to the relative operation and maintenance costs for these two types of systems.

Grinder pump pressure sewer systems present further disadvantages due to the solids and grease involved. Provisions must be made to overcome the problems involved due to the collection of solids and grease. These provisions should include flushing stations to remove solids and grease buildup from the collection system and cleaning and treatment facilities which will remove and treat solids and grease at the wastewater treatment plant. STEP pressure sewer systems have the further potential disadvantages of possible odor problems due to the fact that STEP wastewater is anaerobic in nature and therefore has the potential to produce hydrogen sulfide. The designer should take this into consideration and should make certain that proper ventilation is provided at each pressurization unit. In addition, odor removal methods may need to be considered at air release valves, at pump stations, and at any discharge to atmosphere.

One further disadvantage of STEP systems is that the designer must also consider septage disposal. Septage is defined as the treated waste solids which accumulate in the interceptor tank. These solids must eventually be pumped out of the interceptor tank and subsequently transferred and treated in an acceptable manner. This is covered in more detail in Section 4.2.1-Septage Removal.

3.3   OTHER CONSIDERATIONS

At this time, pressure sewers do not have an extremely long history which would provide data and experience upon which to rely. This is in direct contrast to conventional gravity sewer systems. Consequently, the design of pressure sewer systems will not be as established as conventional gravity sewer design until this history and experience is established. The hydraulics of pressure sewer system design will be the most affected by the lack of operating experience, especially considering the fact that little data is available on the discharge of numerous pumps into a common main. This is also complicated by the fact that the pumps are located at different elevations and at different locations along the main.

Air entrapment and entrainment within pressure sewer mains can be a very big concern. Where the lines empty by gravity flow during non-use times and fill with air, special analysis is required to account for headlosses and other detrimental flow conditions presented by the presence of this air.

Since pressure sewer mains are generally much closer to the ground surface than conventional gravity sewer mains, the pipes have a greater potential to be damaged by excavating equipment. Proper precautions should be taken to help minimize possible damage to the pipe. These precautions may include marking tape buried above the pipe, wire or tape buried with the pipe, color coding the pipe for easy identification, and maintaining an accurate up-to-date set of system maps.

4
Preliminary Design Concepts and Considerations

4.1 SELECTION OF THE TYPE OF PRESSURE SEWER SYSTEM

As stated earlier, there are two types of pressure sewers, grinder pump systems and STEP systems. Table 4-1 has been taken from a chart that has been prepared to assist in the choice between grinder pump systems and STEP Systems.

The major capital and operation and maintenance costs of pressure sewers have historically related to the on-lot or pressurization facilities. Several considerations must be factored into on-lot pressure sewer system design. These include:

1) The type of pressurization unit.
2) Whether or not single or multiple service pressurization units are utilized.
3) Location of pressurization units.
4) Alarms and controls.
5) Aesthetics and safety problems.
6) Serviceability of components.
7) Materials of construction.

28 PRESSURE SEWER SYSTEMS

Table 4-1. A Comparison of the Relative Costs of Grinder Pump and STEP Pressure Sewer Systems[13]

Item of Comparison	Grinder Pump Systems	STEP Systems
Capital cost: on-lot		
pressurization unit	more	less
appurtenances	less	more
Capital cost: main	similar	similar
O/M cost on-lot		
pressurization unit	more	less
residuals handling	less	more
O/M cost: main		
$\frac{present\ population}{design\ population} = 1$	similar	similar
$\frac{present\ population}{design\ population} \ll 1$	more	less
Treatment Plant		
capital cost	more	less
O/M cost	more	less
Hydrogen Sulfide and Odor Potential	less	more

8) Electrical problems.
9) Contingency problems.

The first on-lot facility design decision is the specification of the type of pressurization unit. This will have specific effects on the remainder of the system design. Unless local circumstances preclude one of the two primary alternatives, both the grinder pump and septic tank effluent pump (STEP) systems should be considered. Without a concentrated feasibility study for each individual project, it cannot be categorically stated that either of these types

PRELIMINARY DESIGN CONCEPTS 29

of pressure sewers is superior to the other. Kreissl[4] recommends that the relative merits of each system be weighed carefully by the engineer in his evaluation of cost effectiveness.

Bowne[10] offers the following advantages of STEP Systems over grinder pump systems:

1. Grinder pumps have a history of mechanical failures and clogging. Effluent pumps, on the other hand, when properly used, have a good performance record.
2. The grease and macerated solids passed by the grinder pumps tend to clog the pipe, making hydraulic design difficult, as cleansing velocities must be maintained in the pipelines. Interceptor tanks remove 70 to 90 percent of the grease and all of the grit. Only very small, easily suspended solids are passed and these pose no problem to the pipe system hydraulics.
3. Within the space provided between the liquid level in an interceptor tank and the roof of the tank, more than one day's sewage may be stored. This means there is a buffer in the event of power outage or equipment failure, and an ability to temper the system's peaking factor.
4. Interceptor tanks and effluent pumps are generally less expensive than grinder pump installations.
5. Some existing septic tanks can be used. The number is small, though, since in areas of extremely high groundwater many tanks leak, thus providing infiltration, an intolerable situation.
6. Some pretreatment is accomplished in the interceptor tank. The septic tank effluent is therefore free of grit and possesses substantially reduced biochemical oxygen demand (BOD) and suspended solids (SS). At least 50 percent reduction in BOD and SS can be expected following septic tank treatment.
7. At the treatment plant, bar screens, comminuters, grit chambers, and the primary clarifier may be

30 PRESSURE SEWER SYSTEMS

eliminated owing to the lesser flows and reduced BOD and SS levels to be treated.

Bowne also offers the following disadvantage of STEP systems[10]:

The interceptor or septic tanks must be pumped and the contents disposed. This is not thought to be a major problem, but the cost of pumping and disposal must be taken into account when comparing this type of system with grinder pump systems.

In addition, Kreissl[4] has pointed out that when the population served at the time of construction is significantly less than the projected maximum population, the hydraulic design for the pressure sewer system may be difficult, especially for grinder pump pressure sewer systems where critical scouring velocities are required.

4.2 CONSIDERATIONS IN THE SELECTION OF A STEP SYSTEM

The STEP system pressurization unit consists of a septic tank or interceptor tank, pump chamber, pump, and control system (as shown in Figure 3-2). Some equipment suppliers are packaging STEP systems; however, most designers have chosen to build the system using various specified components. Wastewater from the home flows by gravity to the septic tank. Rather than use the word septic due to its negative connotation, many designers have chosen to call the septic tank an interceptor tank. In the interceptor tank, anaerobic biological degradation of organic materials occurs in addition to the settling of solids and grease accumulation. Gases from this anaerobic action are vented through the house plumbing to roof vents.

If properly designed, the interceptor tank can remove between 80–90% of the grease, 70–90% of suspended sol-

PRELIMINARY DESIGN CONCEPTS 31

ids including all of the grit, and 50–80% of the BOD.[20] Tank materials being used today include concrete, coated steel, fiberglass, and polyethelene plastic. Existing septic tanks are sometimes used; however, they should be checked for inlet and outlet configuration and leakage. In one project in Oregon, 500 existing tanks were checked, and fewer than 10% were usable.

The pump chamber can be integral to or follow after the interceptor tank as a separate chamber. Liquid from the interceptor tank flows to the pump chamber containing the pump and level controls. The level controls are set to operate within a specific range of liquid volume. Control systems in common use include mercury floats, pressure sensing switches, diaphragm switches, and magnetic weight displacement switches. It should be noted that mercury float switches have proven to be very successful and very reliable, and they are used much more widely than the other means of level controllers. A single mercury float is often used to actuate a high-level alarm system.

The most commonly used pumps in STEP systems are the small centrifugal submersible effluent pumps. These pumps range in size from 1/4 HP to 2 HP and are capable of passing 1/4" to 1" solids.

4.2.1 Septage Removal

The major disadvantage of using a STEP System is that the septage (grease and sludge) must eventually be pumped out of the interceptor tank. However, by eliminating the soil absorption system with its potential backflow problems, pumping should not be required for at least ten (10) years. There is a great deal of data available which would indicate that the sludge buildup in a septic tank or interceptor tank should be on the average of ten gallons per capita per year.[9]

Cooper and Rezek[67] have reported that, in general, experience at field sites they visited has indicated that septic tank pumping can occur at ten-year intervals. Periodic checks dating back to 1970 at both Port Charlotte and Port St. Lucie, Florida, reveal that pumping at ten-year periods should be adequate. Bowne, in Oregon, agrees that such an interval should be adequate, provided that yearly inspections at individual sites justify this generalization. At the Coolin and Kalispell sewer districts in Idaho, current maintenance schedules call for septic tank pumping at four-year intervals; however, experience to date indicates that longer intervals would be acceptable.

A comprehensive study of septic tanks has been performed by the Department of Health, Education, and Welfare.[54,59] Over 600 references were studied to develop information on previous research, and practice and experience were reviewed in 12 countries plus the United States. Over 200 tanks in operation were studied along with many laboratory models. That study shows, for example, that a 900-gallon septic tank serving the average home would have to be pumped every 25 years. However, as the sludge and scum layers build, less detention time is developed so a 25-year pumping interval is longer than recommended. When used in conjunction with drainfields where even the smallest discharge of solids is undesirable, the tanks should be pumped more frequently, but not as frequently as many people believe. Overly frequent pumping is a practice done mostly in hopes of curing failing drainfields.[10]

In 1949, a study by the U.S. Public Health Service showed that septic tank sludge accumulates at a rate of ten gallons per capita per year through the first six years and slightly less thereafter.[9] Measurements by Bowne[76] at the Glide Pressure Sewer System have verified this data and Bowne has estimated that the frequency of septage pumping is expected to average ten or more years.

Figure 4-1 projects the cleanout rate of various size septic tanks with productive soil absorption systems as produced by Langford.[9]

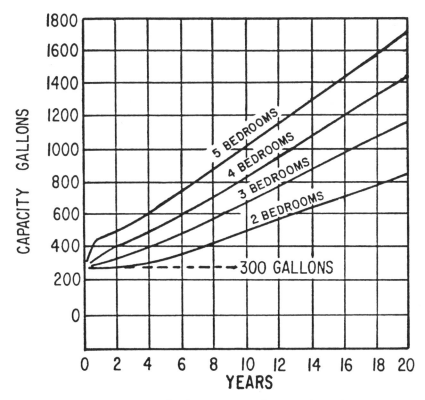

Figure 4-1. Anticipated frequency of pumping interceptor tanks.

Cooper and Rezek[63] have prepared a report for the EPA on septage treatment and disposal, and their findings indicate that treatment will not be a major problem. However, it will certainly have to be taken into account in the design of a STEP pressure sewer system.

4.2.2 Other Considerations for STEP Systems

Since an anaerobic septic tank generates methane and hydrogen sulfide, the venting system must be adequate to dispose of the gases that are generated. The standard house plumbing venting system used with an onsite disposal system will perform the same function with an interceptor tank. In addition, Bowne[75] has recommended that soil vent fields be used to asist the venting within STEP systems, although he also recommends that the designer be very careful concerning infiltration of groundwater. It is also recommended that intermediate pumping stations and air release valves in an effluent pressure sewer system be vented to an underground drainfield if the pumping station is located near residences.

Since septic effluent can be more corrosive than fresh sewage, especially if the septic effluent is exposed to air, it is recommended that materials making up the pumping system be as noncorrosive as possible. The preferred materials for piping and valves is plastic or bronze.[67]

Of growing importance is the power cost to operate the pumping unit and how the pump would perform in potential brownout situations. If a homeowner can use an effluent pump of 1/2 horsepower or less rather than a 2 horsepower centrifugal grinder pump, the power cost would be reduced. This is not a significant amount now, but higher electric costs to the homeowner will become more and more important in the future.[9]

4.3 GRINDER PUMP SYSTEMS

The grinder pump system pressurization unit consists of a pump chamber, pump, and controls (as shown in Figure 3-1). Grinder pump systems are available in packages and most designers prefer to use packaged systems for simplicity. These packages include check valves, level controls, gate valves, and control panels. This chamber can be installed in the basement, as is often done in northern climates.

4.4 STORAGE VOLUME

Due to potential power outages, both STEP and grinder pump installations should have reserve holding capacity above the normal operating volume. Most designers are overly concerned when considering the duration of most power outages. Results of a study compiled by the Federal Power Commission on 187 power outages which occurred from 1968 to 1972 are tabulated below[20]:

Generally speaking, grinder pump systems have approximately 50 gallons of storage whereas STEP systems have from 150 to 200 gallons of storage due to the configuration of the interceptor tank.

Table 4-2. Duration of Power Outages as Reported by the Federal Power Commission

% of Total Outages	Duration
53%	<1 Hr
81	<2 Hr
89	<3 Hr
95	<5 Hr
97	<9 Hr

4.5 A COMPARISON OF SEMI-POSITIVE DISPLACEMENT PUMPS WITH CENTRIFUGAL PUMPS

Two types of pumps are available for use in grinder pump systems. They are the semi-positive displacement (SPD) pump and the submersible centrifugal pump. Environment One Corporation is the only manufacturer of the SPD pump, and this pump has been shown capable of operating above 80 feet of head. The pump characteristic curve is nearly vertical which means that the pump will discharge about the same amount of flow no matter what the discharge pressure. This may tend to simplify the hydraulic design of the piping system due to the predictable discharge of each pump. The extreme operating condition for this type of grinder pump system would occur after a power outage when all pumps would come on line at once. It is possible for this pump to discharge at destructive pressures, however, thermal overload protectors are included in an attempt to prevent damage to the pump motor and the system piping.

Centrifugal grinder pumps are manufactured by many companies, examples of which are Hydromatic, Peabody-Barnes, Myers, and Toran. Centrifugal pumps have a self regulating head-discharge curve which desirably enables the pump to deliver high flows during times when few pumps are operating. This means that scouring velocities, which are critical in grinder pump systems, will be achieved frequently within systems which utilize centrifugal pumps. Also, centrifugal pumps can operate at shutoff head, which is a high pressure-no discharge situation, without damage. Positive displacement pumps, on the other hand, are more expensive, more difficult to repair, and rely on mechanical devices to prevent damage should high pressure occur.[10]

There is one factor which can and should be considered with respect to peak flows within pressure sewer systems. By using centrifugal pumps, peak flows in excess of the design peaks can be accommodated within the system because a centrifugal pump can run at shutoff head. This allows centrifugal pumps to "wait" on one another during peak flow conditions. For example, if a pump comes on and the discharge pressure is greater than shutoff head, the pump will run without discharging until pressures reduce below the shutoff head of the pump. If the system is designed properly, this period of time without discharge will

PRELIMINARY DESIGN CONCEPTS 37

be brief and will not harm the pump in any way. Therefore, the fact that there is a possibility that the design criteria might be exceeded in certain circumstances should create no problems for pressure sewer systems with centrifugal pumps.

In fact, it is recommended that centrifugal pump pressure sewer systems take advantage of the fact that the pumps can wait on each other and thus dampen peak flows. The storage capacity available within the pressurization unit not only provides an emergency storage capability in the event of any type of service interruption, but it could also help to smooth the diurnal peaks which are inherent in sewage flow. By creating greater headloss and therefore greater discharge pressures, the design of smaller pipe sizes will safely suppress peak flows within the collection system itself. This suppression of peak flows will create obvious advantages at the wastewater treatment facility due to the fact that it will tend to reduce capital costs for provisions of flow equalization requirements as well as reduce operating costs because a more consistent hydraulic and organic loading will be placed on the plant. Future design criteria may even further reduce pipe sizing requirements such that all peak flows are virtually eliminated.[25]

Thus, the main advantage of the centrifugal pump over the SPD pump is the inherent characteristics of the pump head versus capacity curve as shown on Figure 4-2. Centrifugal curves have dramatic increases in flow as head decreases, therefore, scouring velocities can be achieved more frequently. Plus, the ability of centrifugal pumps to reach shutoff heads at higher pressures allows them to wait on one another to discharge, and this helps to control peak flows in the piping system.

The advantage that SPD pumps have is that the hydraulic design of the piping system is simpler since the discharge from each pump is more predictable. In addition, it has been reported that SPD pumps are better able to

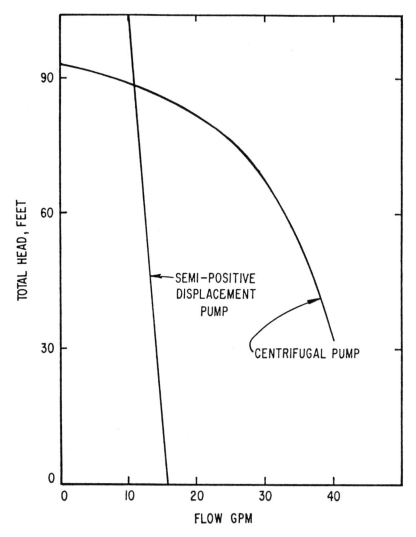

Figure 4-2. Comparison of pump performance curves for semi-positive displacement pumps and centrifugal pumps.

handle air entrapment and grease plugs due to their higher operating pressures.[20]

5
Final Design Considerations

5.1 DESIGN FLOWS

Determination of design flows and the resulting hydraulic analysis are fundamental to the design of a pressure sewer system. At the current time, two general methods exist whereby design flows are established for pressure sewer systems. The first method utilized was developed for semi-positive displacement pumps which pumped very close to the same flow at any given head condition. This method of design is based on the probability of simultaneous pump operation, and it is therefore termed the probability method. The second design method was developed for centrifugal pump pressure sewer systems, and this method is based upon the anticipated wastewater generation from the number of homes served. This method has subsequently been called the Rational Method.[24] In addition, computer programs have been developed that solve the complex calculations of simultaneous pump operations. Even so, these programs have generally only been utilized as design aides due to the difficulty in predicting exactly which pumps will be in operation during the design phase. It is therefore recommended that these computer programs be utilized to refine initial designs as developed by the selected design methodology.

It is recommended that the designer of a pressure sewer system not only carefully research his method of design, but that he also carefully research all factors that could contribute to the design flow. These factors include demographics, housing absorption, and water conservation features, and they will be defined and discussed in more detail in this section.

The following is a more detailed discussion of the two general types of design methodologies along with a further description of the factors that affect the design flow.

5.1.1 Probability Method

This method examines the probability of the number of pumps operating simultaneously in a total population of pumps, and the method has therefore been given the name "Probability Method." Once the probability analysis provides the maximum number of pumps operating simultaneously, one only has to multiply the number of pumps operating by the flow per pump to obtain the maximum design flow. This of course relies on the assumption that the flow per pump is relatively constant, which is a valid assumption in the case of semi-positive displacement pumps.

Table 5-1 is one example of the basis of design for the Probability Method. This table has been provided by the Environment One Corporation which manufactures semi-positive displacement grinder pumps. As an example, suppose a subdivision containing 200 lots were proposed to be served. Assuming that one grinder pump would be located on each lot, inspection of Table 5-1 would show that the maximum number of grinder pumps operating simultaneously would be 11. When an average flow of 11 gallons per minute per grinder pump is used, this means that the maximum or design flow from the subdivision would be 121 gallons per minute. The assumption of 11 gallons per

Table 5-1. Maximum Number of Grinder Pump Cores Operating Simultaneously[17]

Number of Grinder Pump Cores Connected	Maximum Daily Number of Grinder Pump Cores Operating Simultaneously
1	1
2– 3	2
4– 9	3
10– 18	4
19– 30	5
31– 50	6
51– 80	7
81–113	8
114–146	9
147–179	10
180–212	11
213–245	12
246–278	13
279–311	14
312–344	15

minute per pump is valid, and the Probability Method of design has proven to be very successful for the Environment One Pump.

Langford has stated that the probability method is not recommended for centrifugal pumps since the average flow from the pump cannot be predicted due to the characteristic performance curve of the centrifugal pump.[9] Bowne[23] agreed with Langford and stated that, while the assumption of a specific rate of flow delivered by each pump may apply to a semi-positive displacement pump, it does not apply to centrifugal pumps.

On at least one occasion, the Probability Method was utilized to establish design flows for a centrifugal pump pressure sewer system at Hot Springs Village, Arkansas.[22] As stated before, the Probability Method of analysis works well for positive displacement pumps. However, its use with centrifugal pumps has been discouraged.

Further research into the centrifugal pump pressure sewer system has indicated that the Probability Method is not as accurate as would normally be desired. Plus, a new set of design criteria has been published for Hot Springs Village that replaces the criteria utilizing the probability method.[26] The new criteria utilizes the Rational Method as explained in the following text.

5.1.2 Rational Method

In the design of the Glide Pressure Sewer System in Douglas County, Oregon, Bowne has used a method of determining design flows that was developed by the Battelle Institute.[16] Basically, in using this method the pressure sewer mains are sized to accommodate the actual amount of sewage generated by the number of homes served. This method has been termed the "Rational Method" and is preferred for use with centrifugal pump pressure sewer systems. Bowne has stated that, since the Rational Method has been successfully employed before for centrifugal pump pressure sewer systems, and since it is logical, it is his opinion that the rational design approach should be used for these systems.[23]

Figure 5-1 shows a typical design chart taken from the Battelle Institute's study[16] while Figure 5-2 shows a pumped effluent design chart developed by Cooper Communities, Inc. specifically for Tellico Village, Tennessee.[27] Figure 5-3 is a graph of recommended design flows as published in the State of Florida Guidelines.[13]

FINAL DESIGN CONSIDERATIONS 43

No. of Dwellings	Total Av. Flow (GPM)	GPM Av. Flow (GPM)	Peak to Av. Flow Ratio	Peak Flow (GPM)	Suggested Design Flow (GPM)
1	150	.104	4.47	.465	15
5	750	.521	4.40	2.29	25
10	1,500	1.04	4.36	4.53	
20	3,000	2.08	4.30	8.94	35
30	4,500	3.12	4.26	13.29	
40	6,000	4.17	4.22	17.60	
50	7,500	5.21	4.19	21.83	45
60	9,000	6.25	4.17	26.06	
70	10,500	7.29	4.14	30.18	
80	12,000	8.33	4.12	34.32	50
90	13,500	9.37	4.10	38.42	
100	15,000	10.42	4.08	42.51	50
110	16,500	11.46	4.06	46.53	
120	18,000	12.50	4.04	50.50	60
130	19,500	13.54	4.03	54.57	
140	21,000	14.58	4.01	58.47	65
150	22,500	15.62	4.00	62.48	
160	24,000	16.67	3.98	66.35	70
170	25,500	17.71	3.97	70.31	
180	27,000	18.75	3.96	74.25	80
190	28,500	19.79	3.95	78.18	
200	30,000	20.83	3.93	81.86	90
210	31,500	21.87	3.92	85.73	
220	33,000	22.92	3.91	89.62	
230	34,500	23.96	3.90	93.44	100
240	36,000	24.99	3.89	97.21	
250	37,500	26.04	3.88	101.04	
260	39,000	27.08	3.87	104.80	110
270	40,500	28.12	3.86	108.54	
280	42,000	29.16	3.85	112.27	
290	43,500	30.12	3.84	116.00	
300	45,000	31.25	3.83	119.69	125

The suggested design flows in the last column represent a suggested interpolation at random intervals to facilitate line sizing and hydraulic calculations. In the lower range of accumulative homes (under 100), the influence of the storage-pumping action of the contributing units, when compared to the calculated peak flow, will determine the most acceptable lowest suggested design flow to use.

Figure 5-1. Suggested Design Flows for 150 GPD Average Daily Flow (3.0 People Per Dwelling) (50 GPD Per Capita) from Battelle[16]

No. of Lots	Avg. Flow (GPD)	Peak to Average Ratio	Peak Flow (GPM)	Suggested Design Flow (GPM)	Suggested Pipe Size (In.)	Headloss (Ft/1000Ft) C = 140
1	98	4.46	0.30	15	1.5	23.3
2	196	4.45	0.61	15	1.5	23.3
3	294	4.43	0.91	15	1.5	23.3
4	392	4.42	1.20	20	2	9.8
5	490	4.42	1.50	20	2	9.8
10	980	4.38	2.98	30	2	20.7
15	1,470	4.36	4.45	30	2	20.7
20	1,960	4.34	5.90	30	2	20.7
25	2,450	4.32	7.34	40	3	4.9
30	2,940	4.30	8.78	40	3	4.9
40	3,920	4.27	11.63	40	3	4.9
50	4,900	4.25	14.45	40	3	4.9
60	5,880	4.22	17.25	40	3	4.9
70	6,860	4.20	20.03	40	3	4.9
80	7,840	4.19	22.79	45	3	6.1
90	8,820	4.17	25.53	45	3	6.1
100	9,800	4.15	28.25	50	3	7.4
125	12,250	4.12	35.01	50	3	7.4
150	14,700	4.08	41.68	55	3	8.8
175	17,150	4.05	48.28	55	3	8.8
200	19,600	4.03	54.81	60	3	13.8
250	24,500	3.98	67.70	70	3	17.7
300	29,400	3.94	80.39	90	4	5.4
400	39,200	3.87	105.25	110	4	7.9
500	49,000	3.81	129.52	130	4	10.7
600	58,800	3.75	153.30	160	4	15.7
700	68,600	3.71	176.65	180	4	19.5
800	78,400	3.67	199.61	200	6	3.3
900	88,200	3.63	222.24	230	6	4.3
1000	98,000	3.59	244.55	240	6	4.6
1100	107,800	3.56	266.59	270	6	5.8
1200	117,600	3.53	288.36	290	6	6.6
1300	127,400	3.50	309.88	310	6	7.4
1400	137,200	3.48	331.18	330	6	8.3
1500	147,000	3.45	352.27	360	6	9.8
1600	156,800	3.43	373.15	380	6	10.8
1700	166,600	3.40	393.85	400	6	11.9
1800	176,600	3.38	414.37	420	6	13.0
1900	186,200	3.36	434.72	440	6	14.2

Figure 5-2. STEP Pressure Sewer Design Chart

ASSUMPTIONS: 2.3 persons per dwelling
50 gallons per capita per day
85% absorption factor

FINAL DESIGN CONSIDERATIONS

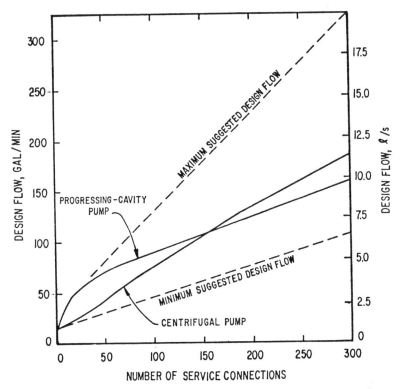

Figure 5-3. Recommended design flow from "Design and Specification Guidelines for Low Pressure Sewer Systems."[13]

Design charts utilized by the Rational Design Method are based upon the following equations.[24]

Qavg = (Persons/Dwelling) × (Flow/Person) × (Dwellings)

Qpeak = P × Qavg

Where:

Qavg is the average flow generated by the given number of dwellings

Qpeak is the design peak flow generated by the given number of dwellings

P is the peak factor as calculated by the following equation

$$P = 1 + \frac{14}{4 + p^{1/2}}$$

where p is the population served in thousands

In addition, these design charts usually contain a suggested design flow which serves a twofold purpose. The first purpose recognizes that for lower numbers of dwellings (i.e., below 100), the influence of the pumps themselves, when compared to the calculated peak flow, will determine a proper design flow. The second purpose of the suggested design flow allows the designer to choose a design flow for a wide range of dwelling numbers and this facilitates hydraulic calculations by making them much simpler.

5.2 DEMOGRAPHIC PROJECTIONS

Demographic projections can have a great deal of importance when designing pressure sewer systems. The first projection normally made by the designer is the number of persons per dwelling, or, as may be more appropriate, persons per connection. Actual census figures are the best to use for this projection. Normal family sizes as measured nationally would tend to average 3.0 to 3.1 persons per dwelling; however, this figure may not hold true for specific types of developments. Cooper Communities, Inc. has successfully utilized a population projection of from 2.3 to 2.5 persons per dwelling within their communities which are retirement and recreational in nature. In addition, it has also been shown that multi-family housing within Cooper Developments has averaged between 1.8 and 2.0 persons per dwelling.[27,28,29] It is recommended that the designer consider this factor carefully before selecting design flows for a pressure sewer system.

FINAL DESIGN CONSIDERATIONS 47

The second projection generally made would be the average wastewater flow per person per day. This projection can be made either by researching literature for similar projects or by actually measuring wastewater flow at the project or at a similar project. In the *Sewerage Study for the Glide-Idleyld Park Area of Douglas County, Oregon*,[11] William Bowne has made a comprehensive literature review of such studies. This review showed an average sewage flow of 47 gallons per capita per day (gpcd). This figure was obtained by taking the mean flow from 19 studies on both new domestic sewage and septic tank effluent.

The highest sewage flow reported in the above mentioned literature review was 78 gpcd. Section 8.0-Characteristics of Pressure Sewage, provides a more detailed discussion of this parameter.

It is widely recognized that wastewater generation is a function of water demand and water demand in the U.S. is commonly estimated to be 100 gpcd, however, this figure includes industrial uses which may or may not be present in any given project. Existing Cooper Communities Developments have successfully utilized a conservative design value of 75 gpcd due to the lack of industrial use and due to lower water usage by retired persons.[26,28,29] These design values have been successfully incorporated into the master plans for Cooper Communities Developments and have been accepted by the Arkansas Department of Health and by the Tennessee Department of Health and Environment.

For planning purposes, daily household water usage may be divided into categories of interior use and exterior use. Interior household use includes toilet flushing, laundry, bathing, dishwashing, drinking, and cooking, and it can be assumed that the interior use will contribute in whole to the wastewater system. Outdoor or exterior use primarily consists of lawn watering, none of which will contribute to the wastewater system. Studies to determine effects of water conservation have summarized the normal interior water use rates found by researchers.

Siegrist[30] references himself and four other researchers who have found interior water use rates of 42.5 gpcd, 44.4 gpcd, 41.2 gpcd, 55.12 gpcd, and 63.5 gpcd. Siegrist et al.[31] also reported an average flow of 42.6 gpcd over a 434-day sampling period. They also tabulated interior flows of 41 gpcd, 45 gpcd, 30 to 50 gpcd, and 44.5 gpcd. By examining the literature, an average interior household water use of 46 gpcd is obtained. Consequently, a conservative design value of 50 gpcd can be used for interior water usage. As stated above, nearly all of the exterior water use is for irrigation, and it can be assumed that exterior watering will not affect the average household sewage flow.

Using water conservation devices in the home can significantly affect the interior average daily water use. Flow reductions ranging from 10 percent to 70 percent have been reported by Maddaus, et al.,[32] Siegrist,[30] and Cole and Sharp[33] in residences where water conserving devices have been used. Palmini and Shelton[34] report a 33-gallon per day (gpd) flow reduction in households with toilet dams, a faucet aerator, and a low-flow showerhead or showerflow control. Maddaus and Feuerstein[35] have studied water conservation and conclude that new residences with low-flush toilets, shower flow controls, and faucet flow controls may exhibit flow reductions of 16.6 gpcd. Milne[36] reported a 13% decrease in water consumption through voluntary conservation and user education. Based upon these studies, and upon an average interior usage of 46 gpcd as previously noted, using water-saver toilets, faucet aerators, and 2.75 to 3.0 gpm showerheads could reduce interior residential water usage to approximately 30 gpcd.

Water conservation can be used to lessen the overall cost of a pressure sewer system. However, the designer should be careful to not detract from the intended lifestyle at the development.

One final projection that may have an extremely important effect upon design flows is the housing absorption fac-

FINAL DESIGN CONSIDERATIONS 49

tor. An absorption factor is defined as the percentage of lots that may be built upon within a given development. This is an extremely important factor, especially in the design of pressure sewer systems for retirement and recreational communities. Cooper Communities, Inc. has proposed an absorption factor of 85% for low-density housing at Tellico Village, Tennessee[27] and has successfully utilized an absorption factor of 80% at three existing developments.[26,28,29] This factor may or may not have impact when considering a pressure sewer system for an existing development. In any case, it is recommended that a study be performed to see if an absorption factor should be taken into account.

5.3 IMPORTANCE OF PROPER DESIGN FLOW SELECTION

Whenever pump vaults with a small reserve space are used and especially when positive displacement pumps are used, it is critical to the overall design of the pressure sewer system that the piping system must be capable of accepting a maximum flow possible when some maximum probable number of pumps run simultaneously. This assumption results in the need for comparatively large diameter pipes. At the other extreme for centrifugal pump systems utilized with interceptor tanks which have relatively large reserve capacity, some investigators might say the piping system was adequate if the reserve space is never exhausted, causing the tank to overflow. This condition would be caused by pump domination, and is associated with the use of pipes that are too small.[23] Neither of these extremes seem prudent, although some pump domination can be desirable in order to inhibit peak flows.

In fact, pressure sewer design criteria used by Cooper Communities, Inc.[24] begins to utilize the advantage that can accrue from dampening of peak flows. According to most STEP pressure sewer design criteria, the pump tank specified has a minimum 24-hour storage capacity.[13] This storage capacity not only provides an emergency storage capability in the event of any type of service interruption, but it could also be used as a flow equalization basin that would help to smooth the diurnal peaks which are inherent in sewage flow. Small pipe sizes will begin to safely suppress the peak flows within the collection system itself. This suppression of peak flows will create obvious advantages at the wastewater treatment facility due to the fact that it will tend to reduce capital costs for provisions of flow equalization requirements as well as reduce operating costs because a more consistent hydraulic and organic loading will be placed on the plant. Future design criteria may even further reduce pipe sizing requirements such that all peak flows are virtually eliminated.

When called on to make judgments, one might pose questions as to what effect an error in judgment would have. Suppose, for example, that a pressure sewer is designed using a piping system that is of unnecessarily large diameter. This could be a costly error in terms of initial construction expense, but would seem to err on the side of safety. The proper operation of air release valves and need for pressure sustaining devices would seem less critical. However, there are detrimental effects in addition to increased construction expense from the oversized pipe. The piping system headloss curve would be flatter, and the pumps would discharge at a higher rate of flow than desired. The resulting high rate of flow would then place an unnecessary demand on downstream mainline pump stations, and also on the treatment facilities. Flows would be extremely variable, and this condition is neither desired nor necessary.

With this in mind, one might now consider the effects of smaller diameter pipe. First, suppose the judgment made was only slightly in error, resulting in only slightly undersized pipe. Then, during the brief and occasional peak moments for which the system was designed, the pumps would run longer than had been anticipated, and discharge at a lower rate than planned. This is not considered to be detrimental. If the judgment error had been even greater, some of the pumps may even run at shutoff head intermittently as they would be dominated by others. So long as this is a sufficiently brief and infrequent occurrence, this is still not considered detrimental. This, of course, assumes that the pump vault contained sufficient reserve space to accommodate the flow from the home during the period the pump was being dominated. At the extreme, should a pump run at shutoff head for so long as to cause overheating, or if the reserve space is exhausted, causing overflow, there is no question that failure has occurred. In that case, the piping system would have been very much underdesigned, and could result in the need for very costly pipe replacement or other costly remedial action.

Overheating of the pump requires a very long operation at shutoff head. This is especially true if a small hole has been provided in the discharge fitting which would allow circulation. (This would also help to prevent air binding of the pump.) To gain perspective, calculations can be made based on certain assumptions. Since one horsepower-hour equals 2547 BTU, and if an assumed 1/2 horsepower pump should run at shutoff head in 50 gallons of water for 1 hour, the rise in water temperature would be only 3 degrees. This assumes that the pump is 100% efficient in heating the water. Also, this discussion neglects consideration of the thermal overload built into pressure sewer pumps. Consequently, overheating seems to be a problem that may be dismissed.

The cost of energy consumption may be similarly estimated. Since 1 horsepower equals 0.7457 kilowatts, a 1/2 horsepower pump operating for 1 hour will use 0.4 KWH. At the cost of 5 cents per KWH, the cost for operating the pump for 1 hour would be less than 3 cents. However, this discussion does not suggest that a 1-hour period of running at shutoff is recommended. It was only selected to describe the order of magnitude.

Overflow would be a concern if a small vault were planned. However, standard STEP pressure sewer systems include a 50-gallon reserve between pump-on and alarm level, plus up to a 200-gallon storage between alarm and overflow. This would seem to provide a very substantial safety factor.

A review of these considerations suggests that there are many safeguards inherent in the design of pressure sewer systems employing centrifugal pumps with interceptor tanks which have large reserve capacity.

This review is meant to impress the reader with the importance of properly estimating the design flow and sizing the pipes for a pressure sewer system. Pressure sewers are one area where a conservative (i.e.; higher design flow than could be expected) approach could very well be detrimental to the system.[23]

5.4 HYDRAULICS

Plastic pipe is the most common type of pipe utilized for pressure sewers, and it has been shown to exhibit a Hazen-Williams roughness coefficient (C) in English units of 155 to 160 with clean water. However, due to the nature of wastewater and the potential for grease deposition and microbiological growth on the walls of the pipe, a reduced value of 130 to 150 has been recommended.[13]

A "C" value of 155 is recommended by Neale and Price[15] for smooth plastic pipe. An American Water Works Association committee report[16] on plastic pipe states that a Hazen-Williams coefficient of 160 has been observed in plastic pipe, and a conservative value of 150 can be used in determining flow quantities.[44]

In general, the specific type of pipe used for pressure

FINAL DESIGN CONSIDERATIONS 53

sewer systems will be poly-vinyl chloride (PVC), SDR-26; however, in those situations where the pipe would be subjected to undue stresses, pipes with greater pressure rating should be specified.[27] A Hazen-Williams C-factor of 140 has been successfully utilized in the design of pumped effluent systems in Arkansas and is accepted by the Arkansas Department of Health.[27]

Kreissl[4] has recommended that care be taken in the design of pressure sewer systems so that the piping is never mistaken for a water system. Many water systems utilize the same types of pipe that have been recommended for pressure sewer systems, and this would create the possibility of a mistaken water tap onto a pressure sewer main unless preventive measures are taken. Some pressure sewer systems have encouraged strict color coding of pressure sewer and water pipes to prevent this mistake from happening. An example would be to require all potable water pipe to be blue in color and all pressure sewer pipe to be white in color.

The hydraulic design of a pressure sewer must take into account several factors, the most noteworthy being the head-discharge characteristics of the pressurization unit. The simplest case is that of a centrifugal STEP system. Pipe sizes should be selected which display the best combination of low frictional headloss and reasonable velocity at design flow. Most pressure sewer systems will employ pipe sizes of increasing diameter when progressing from the origin toward the terminus of the system. It is recommended that a centrifugal pump should not be specified under conditions requiring greater than 85% of the available head when operating alone.[9,13]

The following procedure is typically used to approximate the initial hydraulic design[13]:

1. Determine the ultimate number of facilities to be served by the system.
2. Choose a design flow from a design chart prepared specifically for the project.

3. Prepare a plan and profile of the proposed system.
4. Evaluate the need for air release and pressure sustaining valves.
5. Plot hydraulic grade lines corresponding to various pipe sizes. Any pipe size which indicates an excessive total dynamic head is sequentially discarded until a proper one is found based on economics, pressure limitations, and reasonable approximation of pump characteristics.
6. Prepare a dynamic hydraulic grade line based on previous determinations. Individual pump units can then be selected based on site-specific head conditions and desired flow rate. Individual pump characteristics can be tested for sufficiency by checking the elevation difference between the pump and the dynamic head where the pump lateral intersects the mainline.

The initial hydraulic design is then tested in the following manner:

1. Plot the system head curve of the pump (incuding losses in service lines and fittings).
2. Locate head requirements at design flow and determine adequacy of pump and suitable pipe size.

It is at this point that a computer program similar to the one developed by the F. E. Myers Company[19] becomes useful, especially for larger systems.

The pressurization unit selection will depend upon the hydraulic profile of the system and the characteristic pump curves chosen for the system. Thus, an analysis of the manifolded pump and pipe networks should be determined for the proposed system. An analysis of the time dependent alternations in the manifold system's characteristics as pumps turn on and off should be included to determine the proposed system capabilities. Water hammer and surge analyses may be necessary on large systems with

FINAL DESIGN CONSIDERATIONS 55

higher pressures but are not normally a concern. Factors to be taken into consideration in performing the analysis include[13]:

1. Operating capacity of the pump chamber or wetwell.
2. Pump characteristics.
3. Distribution piping, materials, and appurtenances.

Design procedures are discussed in more detail in Section 6.0-Design Methodology wherein a specific design is shown for a theoretical project.

It should be noted here that within a STEP system, not all connections will necessarily require a pressurization unit. If the hydraulics can be designed properly, it is very possible and even desirable to combine small diameter gravity sewers with a STEP pressure sewer system. This design process has been successfully utilized by Bowne[68] at the Glide, Oregon Pressure Sewer System and by Cooper Communities, Inc.[65] at the Metfield Pressure Sewer System in Bella Vista Village, Arkansas. In addition, the Metfield Pressure Sewer System contains one connection which is served by a dosing siphon. Care should be taken within such systems to make absolutely certain that the gravity flow units are well above any expected system hydraulic gradelines.

5.4.1 Minimum Scouring Velocity

The aforementioned discussion of hydraulics will provide one with the information to size the piping system for a pressure sewer based upon the hydraulics involved. However, this is often not the only consideration, especially within grinder pump pressure sewers. Grinder pump pressure sewer mains require sufficient liquid velocity within the pipe to scour grease and solids from the pipe.

The minimum transport velocity for use in designing

56 PRESSURE SEWER SYSTEMS

hypothetical pressurized sewage systems, according to McPherson,[49] is:

$V = D/2$ where V is in feet per second (fps)

D = internal pipe diameter in inches

Fair and Geyer[50] state that self-cleaning velocities of 2-2.5 fps are necessary in order to hold in check the fouling of sanitary sewers by the deposition of waste matter. They also state that fine sand is ordinarily transported by water at velocities of 1 fps or more and gravel at 2 fps or more.

In Recommended Standards for Sewage Works,[48] it is stated that sewers shall be designed and constructed to give, when flowing full, mean velocities of not less than 2 fps.

Several literature sources recommend values for scouring or self-cleaning velocities when grease accumulation is a problem. Those sources recommend a minimum velocity of 2.0-2.5 fps, and they were generally prepared for grinder pump pressure sewer systems.

Hendricks[52] indicates that the maintenance of a scouring velocity for grease control can be a problem if the design line capacities are set by peak sewage flows based on excessively high peak-to-average flow ratios.

For septic tank effluents with greatly reduced solids and grease concentrations, no peak velocity requirements have been determined, but it is recommended that the peak velocity be at least 1 fps to insure the scouring of any suspended material.[13] This is considered as a major advantage for STEP systems since many pressure sewer systems are proposed for areas of low initial capacity or seasonal use where it would be difficult to provide strict flushing velocities at all times.

5.7 CONSIDERATIONS FOR AIR ENTRAPMENT WITHIN THE PIPE NETWORK

Several investigators[10,16,68] have recommended positive pressure maintenance within a system, and this would preclude considerations for air entrapment. However, Kreissl[4] recommends that combined pressure and gravity systems be considered where feasible.

Initially, Bowne recommended that the designer of a pressure sewer system consider the use of pressure sustaining valves in downhill pumping instances so that all air could be excluded from the pipe.[10] The Guidelines from the State of Florida[13] tend to agree with those recommendations. Bowne was subsequently not able to find a pressure sustaining valve suitable for the intended use.[14,78] Consequently, state-of-the-art for pressure sewers has been to design the system for air entrapment and to allow sufficient air and vacuum escape points within the system.

Bowne[78] has stated that STEP pressure sewer systems can be engineered to flow downhill and avoid air problems. He recommends that air release stations be placed a short distance downstream from the absolute high points within the system in order to allow air expulsion during higher flows. He also recommends that standpipes be considered in order to adjust hydraulics for maximum air removal. Bowne noted the disadvantages of grease and solids in downhill pumping situations and strongly discouraged the use of grinder pumps if a downhill pumping situation must be overcome.

Preferably, pressure sewer systems should be oriented such that flow is in the upslope direction, and such that the outfall is at a higher elevation than any significant portion of the collection system. However, this is not always possible since many pressure sewer systems will require downslope pumping. This condition will cause large quantities of air to enter the main, and this can result in hydraulic difficulties such as high headlosses, irregularity of flow, and unpredictability of flow. In the study of this condition, Bowne has recommended that reference should be made to work by Kent,[39] Burton and Nelson,[40] Whitsett,[41] and Winn.[42]

If air is not expelled from a downsloping pipeline, headlosses will be encountered due to the upward component of the air opposing the downward component of the flow. In one example from Kent,[39] for downward pipe slope of 5%, and an air content of 20%, an additional headloss of 10 feet per thousand feet would result. Frequently, when pumping downslope these losses are too excessive to permit, though each design should be considered separately.

In any pressurized closed-conduit system such as a pressure sewer system, the effects of air within the lines must be recognized. For example, an air pocket will decrease the cross-sectional area. Also, the frictional pressure loss for two-phase flow is always greater than the pressure loss for each phase flowing alone at its respective mass-flow rate. Therefore, air in the line increases the system flow resistance, which in turn increases the head against which the pumps in the system must operate.

Recommended Standards for Sewage Works[48] states that automatic air-relief valves should be placed at high points in forcemains to prevent air locking. Lescovich has recommended that these high points be referenced to the hydraulic gradient and not to a horizontal datum line. In addition, Lescovich has recommended that air release or air and vacuum release valves be placed at 1500-foot to 3000-foot intervals along most pipelines.[47]

Air in a piping system will tend to collect at high points in the line when flow velocities are low. If the air pocket thus formed is large, part of it will be removed when the velocity increases. This partitioned air may or may not go through the system, depending on velocity, pipe size, and downslope. It may only move into the sloping straight

FINAL DESIGN CONSIDERATIONS 59

section and then return to the summit when the velocity decreases.

At the velocities expected in pressure sewer systems, not all of the air at a summit will be removed. Air in a downward-sloping pipe tends to collect into a single bubble, which remains stationary when the buoyant forces balance the drag forces. The velocity at which this balance occurs is a function of the line size and the slope. For example, in a 4-inch line with a bubble greater than 6 inches long (1.5 times the line size), the equilibrium velocity is 2.7 fps for a 15-degree slope and 3.4 fps for a 60-degree slope. In a 6-inch line the velocities would be 3.8 and 4.8 fps, respectively. A velocity of 1 fps will remove trapped air from 1½-inch lines if the downslope is less than 10 degrees and from 3-inch lines if the downslope is less than 5 degrees.[44] Thus it is recommended that, as a minimum, air and vacuum relief devices be provided at a short distance downstream from the summits of lines having greater downslopes than those described, however, it should be noted that exact criteria for the need and placement of air and vacuum release valves are lacking.[4]

With higher flow rates, minimum downstream slopes and short travel distances to subsequent low points, the need for air release valves is marginal. The need for air release valves should be closely examined for any downslopes in excess of ten percent (10%). Locations with lesser slopes, where long downstream pipe volume is in excess of that which would be expected to be pumped during one continuous pumping interval, may also require air release valves.

Adequate preventive measures should be taken to avoid the accumulation of gases and air in pressure sewer mains.[13] These include:

1. Sufficient purging after pressure sewer main filling and testing.
2. Submersion of the pressurization unit pump intake to prevent siphoning or vortexing after shutoff.

3. Proper design to prevent undue retention time of wastes in pressure sewers due to biological and chemical activity which may produce gases.

In summary, little is known about the effects of air in pressure sewers, and further research has been requested. Current methods of design usually allow air to freely enter the pipe, thus creating a non-gradient gravity flow situation in downhill pipe sections. Care should be taken to introduce additional headloss factors into these downhill sections to account for overcoming the entrapped air. Care should also be taken to ensure that air and vacuum release valves are properly placed within such a system and to prevent odors at air release points within a pressure sewer system.

6
Design Methodology

As stated previously, there are two general methods utilized for the design of pressure sewer systems. The Rational Method has been generally employed in the design of systems utilizing centrifugal pumps, and the Probability Method has been employed for those systems which utilize semi-positive displacement pumps. Since the vast majority of new pressure sewer systems utilize centrifugal pumps, the Rational Method will be explained in more detail, and brief comparisons will then be made to the Probability Method.

In order to better understand the design of a pressure sewer system, it is usually advantageous to compare such a system with a standard gravity sewer design for the same project. Indeed, an engineer would be remiss if he did not make such a comparison for any project for which pressure sewers are proposed. For our purposes, a "typical" subdivision as shown in Figure 6-1 should provide us with a simple comparison which will serve to illustrate an example of design procedure. Figure 6-1 shows a gravity sewer serving the subdivision in accordance with generally accepted design methodology. In addition, Figure 6-2 shows profiles of the gravity sewer mains designed to serve this subdivision.

Upon review of Figures 6-1 and 6-2, one will note many items that often make gravity sewer systems so expensive.

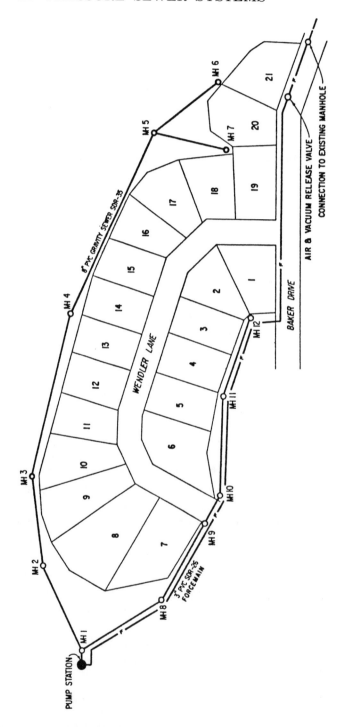

Figure 6-1. Plan view of gravity sewer design for Armstrong Subdivision.

DESIGN METHODOLOGY 63

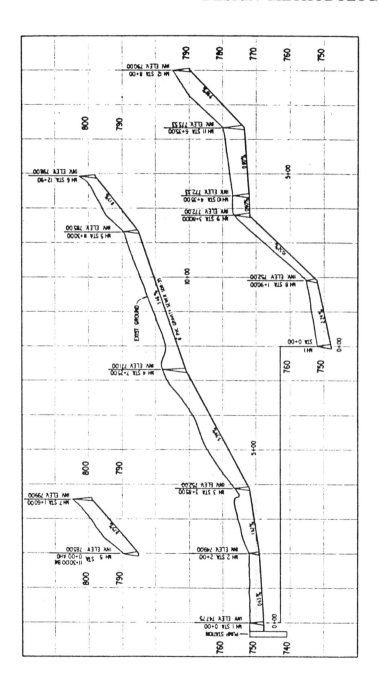

Figure 6-2. Profile of gravity sewer design for Armstrong Subdivision.

64 PRESSURE SEWER SYSTEMS

For example, note that a pump station is required and that the inflexibility of gravity sewer systems requires that mains be placed downhill from the lots, which would require special easements in this particular case. Also note the many manholes required, the minimum size pipe (eight inches), and the trench depth of up to eight feet deep.

Figure 6-3 shows a pressure sewer design for the same subdivision. In order to obtain the pipe sizes as shown on Figure 6-3, a design chart such as the one shown in Figure 5-2 should be consulted. Preliminary pipe sizes can be chosen from a design chart, and further hydraulic calculations will serve to further refine and finalize pipe sizes. While reviewing Figure 6-3, note that no central pump station is required, no manholes are required, and the maximum size pipe is three inches in diameter.

Figure 6-4 shows a profile of the pressure sewer design for the typical subdivision. Note that the trench depth need only be deep enough to prevent freezing problems. Also note that a hydraulic gradient is shown on this profile. This hydraulic gradient can be defined as the pressure of the liquid in the pipe during design flow conditions. One method of calculating this hydraulic gradient, and the significance of the hydraulic gradient will be shown in the following text.

The significance of the hydraulic gradient is that it provides the design engineer with head (or pressure) and flow conditions at any point in the pressure sewer system. Calculation of the hydraulic gradient is sequential in nature and must start at the furthest downstream point in the piping system. Extending the calculations to each terminus of the system will provide design head and flow conditions at each pressurization unit, and this will allow the design engineer to select a pump which will operate during design flow conditions.

It should be noted here that it is very difficult to predict absolute peak flows in a pressure sewer system, and it is definitely not recommended to design the piping system to

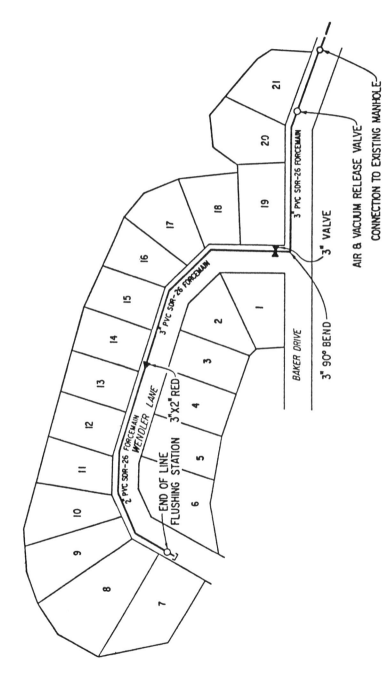

Figure 6-3. Plan view of pressure sewer design for Armstrong Subdivision.

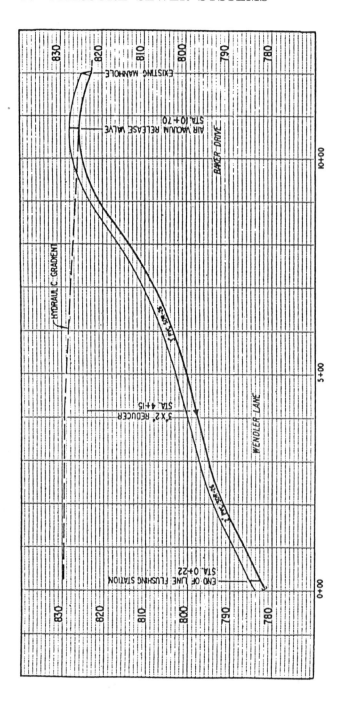

Figure 6-4. Profile of pressure sewer design for Armstrong Subdivision.

allow an absolute peak flow to occur. Pressure sewer systems should be designed with sufficient storage at the pressurization units to allow for a suppression of peak flows which are in excess of the design flow. In addition, since pumped effluent systems generally have much more storage available than grinder pump systems, more suppression of peak flows is possible and is even desirable. Further and more detailed discussions of this subject can be found in Section 5.3-Importance of Proper Design Flow Selection.

Calculation of the hydraulic gradient can best be facilitated on a chart similar to the one shown in Figure 6-5. Careful review of this chart will show that calculated hydraulic gradients should be compared to pipe elevation to see which is higher. In downhill pumping situations, conditions will sometimes occur where the energy gained by flowing downhill is greater than the energy lost through headloss in the pipe. This creates an occurrence known as non-gradient gravity flow, and there will be no pressure in the liquid within the pipe when this condition is encountered. However, the designer is cautioned to consider air entrapment within the pipe when downhill pumping situations are encountered, and more detail is provided on this subject in Section 5.5-Considerations for Air Entrapment Within the Pipe Network.

The specific hydraulic calculations performed for inclusion within Figure 6-5 utilize the Hazen-Williams headloss formula for pressurized water flowing through a closed conduit. The formula as utilized is as follows:

$$S = 0.60 \, R^{-1.167} \left(\frac{V}{C}\right)^{1.852}$$

Where:

S = Headloss in feet per foot length of pipe
R = Hydraulic Radius = $\frac{\text{Area}}{\text{Wetted Perimeter}}$ (Feet)
V = Velocity in feet per second
C = Hazen Williams Constant
   (See Section 5.4-Hydraulics for a further discussion of the Hazen Williams C factor. Figure 6-5 assumes a C of 140)

ARMSTRONG SUBDIVISION PUMPED EFFLUENT DESIGN

*NOTE: PIPE ELEVATION ABOVE CALCULATED HYDRAULIC GRADIENT

STATION	DISTANCE	PIPE SIZE	LOTS SERVED	DESIGN FLOW GPM	HEADLOSS (HL) FT/1000 FT.	TOTAL HL	CALC. HYD. GRADIENT	PIPE ELEVATION	ACT. HYD. GRADIENT	COMMENTS
12 + 00			21						820.00	CONNECTION TO EXISTING MANHOLE
10 + 70	130	3"	20	35	3.83	0.50	820.50	825.75	*825.75	AIR & VACUUM RELEASE VALVE
4 + 15	655	3"	11	25	2.05	1.35	827.10	798.00	827.10	3" X 2" REDUCER
0 + 22	393	2"	3	15	5.74	2.26	829.36	783.00	829.36	END OF LINE FLUSHING STATION
0 + 00	22	1.5"	1	15	23.30	0.51	829.87	782.00	829.87	2" CAP

STATION	DISTANCE	PIPE SIZE	LOTS SERVED	DESIGN FLOW GPM	HL FT/1000 FT.	TOTAL HL	CALC. HYD. GRADIENT	PIPE ELEVATION	ACT. HYD. GRADIENT	COMMENTS
4 + 50(0 + 00)							827.02	800.00	827.02	FORCEMAIN CONNECTION
0 + 50	50	1 1/2"	1	15	23.30	1.17	828.19	PUMP ELEV. 780.00	828.19	PUMP TANK
PUMP	DESIGN FLOW-GPM 15		DESIGN HEAD-FEET 48.19	PUMP SELECTED HYDR-O-MATIC 1.0 HP SP 100 AH			FLOW AT DESIGN HEAD = 34 GPM			85% SHUTOFF HEAD = 52.7 FEET

Figure 6-5. Hydraulic gradient calculation and pump selection chart.

DESIGN METHODOLOGY 69

This same formula can be used to calculate a hydraulic gradient in a pressure sewer system utilizing semi-positive displacement pumps. The only difference would be in the selection of design flows which would be accomplished from a chart as shown in Figure 5-1. This chart is based upon the Probability Design Method as discussed in Section 5.1.1-Probability Method. It is recommended that this type of chart only be used for systems utilizing semi-positive displacement pumps since a basic assumption of this method is a constant flow from each pump that is operating, and such a condition would be extremely unlikely in centrifugal pump systems. Once design flows are chosen, the basic design procedure will be the same no matter which type of system is chosen.

After the initial hydraulic gradient has been established, the design engineer should inspect the pressures created by that particular gradient. A steep gradient may create pressures too great for the individual pressurization units, and the pipe sizes may need to be increased in order to lower headloss and consequently lower the hydraulic gradient. Conversely, the designer may wish to decrease pipe sizes. The advantages of decreasing pipe sizes are a decrease in overall cost and an overall capability to control and lower peak flows in the system. It therefore becomes obvious that pipe selection is a trial and error process and should be carefully used by an experienced design engineer who is knowledgeable in hydraulics and in the specific concepts of pressure sewer systems.

Even though an example of pump selection is shown within Figure 6-5, it is recommended that a separate chart be set up for this purpose. A separate calculation will be required for each pressurization unit, and there will usually not be sufficient room for these calculations on the hydraulic gradient calculation worksheet. Once the designer knows the design head and flow for each pressurization unit, he need only to compare this to a composite pump curve. A pump is then chosen that comes the closest to design conditions.

70 PRESSURE SEWER SYSTEMS

Pump selection for semi-positive displacement pressure sewer systems is simpler since these types of pumps will operate at nearly any head or pressure condition. This also allows the designer to specify the same pump for every application. However, in many instances, the various disadvantages of this type of system would seem to outweigh the simplicity of pump selection. For a further discussion, see Section 4.5-A Comparison of Semi-Positive Displacement Pumps with Centrifugal Pumps.

6.1 DESIGN CONSIDERATIONS FOR INTERMEDIATE PUMP STATIONS

For instances where pressure sewers are chosen for extremely hilly terrain, intermediate pump stations may be required in order to overcome higher elevation differentials. Within pressure sewer systems utilizing centrifugal pumps, 150 feet of total head is usually all that the on-lot pressurization units can overcome. Heads greater than this will usually require an intermediate pump station. Obviously, these conditions may warrant the consideration of semi-positive displacement pumps since they can overcome a greater amount of head. However, the designer should carefully weigh all the advantages and disadvantages, including all economic considerations, before selecting any type of system.

When designing an intermediate pump station for a pressure sewer system, one must take into account the type of system. Intermediate pump stations for grinder pump pressure sewer systems must be capable of handling solids. Large grinder pumps have been utilized for this purpose and they tend to work very well. The disadvantage of grinder pumps for use in intermediate pump stations is that they are somewhat limited in flow and pressure (flows up to 180 gallons per minute and pressures up to about 150 feet of head). They are also very inefficient in

DESIGN METHODOLOGY 71

comparison to other types of wastewater pumps due to the grinding process which may not be required in an intermediate pump station. In spite of this, larger grinder pumps are often used because of their desirable head to discharge relationships and because of their compatability with the rest of the system.

Solids-handling submersible pumps are also used successfully for grinder pump pressure sewer systems. These pumps can have extremely high flow characteristics if desired, however, they do have severe limitations for use when high pressures are required.

Since solids-handling requirements are greatly reduced for STEP pressure sewer systems, intermediate pump stations do not require solids-handling characteristics. Larger STEP pumps have been used in these instances along with verticle turbine pumps. The larger STEP pumps, which are currently available up to two horsepower, have desirable flow and pressure relationships; however, they are very limited concerning maximum flow. On the other hand, verticle turbine pumps are virtually unlimited in both discharge flow and pressure, and they are extremely efficient.

One drawback for verticle turbine pumps could be explosion-proof requirements. Even though explosion-proof equipment is not generally required at the individual pressurization units, explosion-proof equipment should be considered at intermediate pump stations. This is not a problem with submersible pumps, but it can be a serious consideration with verticle turbine pumps.

It is recommended that the designer of an intermediate pump station for a pressure sewer system take measures to inhibit peak flows as much as possible. Decreasing the design flow as much as possible and increasing the overflow capacity is one method of attaining this goal. Variable speed pumps could possibly be used, and pumps of differing sizes in sequential operation could also be used.

One advantage of dampening peak flows at intermediate pump stations is that design and operation of connections to the downstream forcemain are simpler. It is sometimes necessary to connect pressure sewer mains and laterals into forcemains from intermediate pump stations, and the design can be difficult due to the wide range of hydraulic conditions. It should be noted that these downstream connections will increase the flow in the forcemain, thus increasing the total headloss. This will affect pumping conditions, and these factors should be taken into account during pump selection.

The actual pump selection process is trial and error in nature. A hydraulic gradient should be established which is based upon design flows. Then a pump should be selected which operates slightly in excess of design flows which would allow a small safety factor. The hydraulic gradient should then be recalculated based upon actual pumping conditions, and the selected pump should then be reevaluated in accordance with the new hydraulic gradient. This procedure should be followed until a pump is found that closely matches design conditions. It is interesting to note here that grinder pumps or solids-handling pumps are sometimes used for intermediate pumps in STEP systems due to their unique flow and pressure relationships.

6.2 RELATIVE COSTS OF PRESSURE SEWER SYSTEMS

It is impossible to make any definitive general statements concerning the relative costs of pressure sewer systems. Each potential sewer project should be economically evaluated as to the method of wastewater collection and all costs, including capital costs and operation and maintenance costs, should be considered.

DESIGN METHODOLOGY 73

In order to show one example of a simple capital cost comparison, Tables 6-1 and 6-2 have been prepared. Table 6-1 shows a gravity sewer construction cost estimate for the sewer system shown in Figures 6-1 and 6-2. Table 6-2 provides a pressure sewer construction cost estimate for the same subdivision as shown in Figures 6-3 and 6-4.

As can be seen by reviewing Tables 6-1 and 6-2, the total cost of the pressure sewer alternative for the chosen subdivision is $16,225.00 less than the gravity sewer alternative. This equates to a capital cost savings of $774.05 per lot. Even more important than this savings is the fact the pressure sewer main can be constructed for a total cost of $4,897.50, which equates to $233.21 per lot. The capital expense for the on-lot pressurization units and the on-lot piping, which amounts to $2,400.00 per lot, will not be required until a house is built on any given lot. This can be a major economic advantage as shown in the economic analyses that follow:

Table 6-1. Gravity Sewer Construction Cost Estimate for Armstrong Subdivision 1986 Dollars (21 Lots)

Item	Unit	Quantity	Unit Price	Item Cost
8" Sewermain 0–4' Deep	L.F.	875	8.00	$ 7,000.00
8" Sewermain 4–6' Deep	L.F.	1075	8.50	$ 9,137.50
8" Sewermain 6–8' Deep	L.F.	300	9.50	$ 2,850.00
Manholes 0–4' Deep	EA.	3	600.00	$ 1,800.00
Manholes 4–6' Deep	EA.	7	700.00	$ 4,900.00
Manholes 6–8' Deep	EA.	2	850.00	$ 1,700.00
4" Service Pipe	L.F.	1575	5.00	$ 7,875.00
6" × 6" × 4" Wye	EA.	13	50.00	$ 650.00
Pump Station	EA.	1	30,000.00	$30,000.00
3" Forcemain	L.F.	1440	3.50	$ 5,040.00
Air & Vacuum Release Valve	EA.	1	600.00	$ 600.00
			TOTAL	$71,552.50
			COST PER LOT =	$ 3,407.26

74 PRESSURE SEWER SYSTEMS

Table 6-2. Pressure Sewer Construction Cost Estimate for Armstrong Subdivision 1986 Dollars (21 Lots)

Item	Unit	Quantity	Unit Price	Item Cost
3" Forcemain	L.F.	825	3.50	$ 2,887.50
2" Forcemain	L.F.	400	2.50	$ 1,000.00
3" Valve	EA.	1	150.00	$ 150.00
3" × 2" Reducer	EA.	1	10.00	$ 10.00
End of Line Flushing Station	EA.	1	200.00	$ 200.00
Air & Vacuum Release Valve	EA.	1	600.00	$ 600.00
3" 90° Bend	EA.	1	50.00	$ 50.00
		SUBTOTAL		$ 4,897.50
		SUBTOTAL COST PER LOT		$ 233.21
Pressurization Units (Includes On-Lot Piping)	EA.	21	2,400.00	$50,400.00
		TOTAL		$55,297.50
		TOTAL COST PER LOT		$ 2,633.21

A logical extension of the economic data shown in Tables 6-3, 6-4, and 6-5 would conclude that pressure sewers become even more economically viable in areas of slower housing growth rates. This can be a huge economic advantage in retirement communities where a considerable period of time can often lapse between the time a lot is sold and the time a house is built upon that lot. In addition, many lots may never be built upon, and this is a definite major economic factor in favor of pressure sewer systems.

The economic analyses as shown in Tables 6-3, 6-4, and 6-5 are only provided to show relative examples of the cost savings that may occur if pressure sewers are utilized. Each different project will have unique applications, and it is recommended that separate economic analyses be performed for each project. In addition, the economic analyses

DESIGN METHODOLOGY 75

Table 6-3. Economic Present Worth Analysis of
Gravity Sewer Cost for Armstrong Subdivision

ASSUMPTIONS

1. Capital Cost Investment at $71,377.50
2. Useful Economic Life of 40 Years
3. O & M Cost of $834.12 Per Year
4. Depreciation Cost of $1,784.44 Per Year
5. Average Yearly Interest Rate of 6%

PRESENT WORTH = $110,776.32

Table 6-4. Economic Present Worth Analysis of
Pressure Sewer Cost for Armstrong Subdivision
Assuming All Capital Cost in First Year

ASSUMPTIONS

1. Capital Cost Investment of $55,297.50
2. Useful Economic Life of 40 Years
3. O & M Cost of $950.04 Per Year
4. Capital Replacement Cost of $630.00 Per Year
5. Depreciation Cost of $1,382.44 Per Year
6. Average Yearly Interest Rate of 6%

PRESENT WORTH = $99,870.97

Table 6-5. Economic Present Worth Analysis of
Pressure Sewer Cost for Armstrong Subdivision
Assuming Delayed Housing Growth

ASSUMPTIONS

1. Capital Cost Investment of $4,897.50
2. Three Houses Build Per Year For A Capital Investment of $7,200 Per Year Over a Seven-Year Period
3. Useful Economic Life of 40 Years
4. Average O & M Cost of $878.79 Per Year
5. Average Capital Replacement Cost of $582.75 Per Year
6. Average Depreciation Cost of $1,278.76 Per Year
7. Average Yearly Interest Rate of 6%

PRESENT WORTH = $88,732.61

76 PRESSURE SEWER SYSTEMS

performed may need to be more detailed than those performed here and such factors as varying interest rates, useful economic lives, capital replacement costs, depreciation costs, and inflation may need to be studied in great detail.

7
Equipment and Material Considerations

7.1 PIPE SYSTEM MATERIALS

In general, plastic pipe has been utilized for almost every pressure sewer system. Plastic pipe provides advantages as noted below:

1) Material costs are relatively inexpensive as compared to other types of pipe.
2) Installation costs are generally less expensive.
3) Plastic pipe resists corrosion extremely well.
4) Plastic pipe is available in many different sizes, including the smaller sizes necessary at the pressurization unit.

The type of plastic pipe chosen can appreciably affect the design and economics of the pressure sewer system. PVC pipe has almost exclusively been employed. Both rubber ring and solvent weld joints can be used for PVC Pipe if properly jointed per ASTM requirements although solvent weld joints are not accepted in many states unless the weld is performed under ideal conditions.

As a brief history of the types of pipe utilized at earlier installations, the pressure sewer mains at Albany were

78 PRESSURE SEWER SYSTEMS

PVC Type I Schedule 40 with PVC fittings, and the joints were solvent welded.[5] At Phoenixville, PVC Type I SDR-26 piping was used with PVC Schedule 40 fittings.[6] At Grandview Lake, PVC SDR-26 pipe (solvent-welded) and PVC fittings were used.[7] SDR-26 PVC pipe with solvent welded joints was also used in the Weatherby Lake, Missouri pressure sewer system.[38]

Flanigan and Cudnik[16] recommend the use of SDR-26 PVC piping in all systems whose pumping heads do not exceed 90 feet. Environment/One recommends, in order, SDR-21, Schedule 40, and SDR-26;[17] however, it should be noted that the Environment/One Pumps can operate at a much higher pressure than other pressure sewer pumps. The pressure rating of SDR-26 pipe is 160 psi and SDR-21 is rated at 200 psi, while Schedule 40 pipe varies with pipe diameter. Two- and three-inch Schedule 40 pipes are rated at 277 and 263 psi, respectively.[16] All pressure ratings are at a temperature of 73 degrees F and are generally reduced at higher temperature, to the extent that PVC is not recommended above 150 degrees F.

The safety factor between the pressure rating of SDR-26 PVC pipe and system design pressures in almost all cases exceeds four. Since the other recommended PVC pipes all have greater pressure ratings, their safety factors are subsequently larger. A safety factor of four is common for water supply systems where water hammer conditions are much more likely, and the safety factor for all PVC pipes discussed therefore appears to be adequate. Due to potential imperfections in manufacture and installation of pressure pipeline components, the ratio of pressure pipe rating to the maximum pressure developed by the pressurization unit should not be less than 2:1.

Although polyvinyl chloride (PVC) pipe has been most widely used for pressure sewers, high-density polyethylene pipe (HDPE) has been employed in some installations. This material has certain advantages and disadvantages when compared to PVC. Longer pipe lengths with fewer joints are available, but special joining techniques (butt-fusion) are required. HDPE is similar to PVC in working pressure and roughness coefficient. Although some use of polybutylene for service lines has been reported, there is no present basis for its evaluation as a mainline pipe.

7.2 PIPE SYSTEM APPURTENANCES

The need for terminal and in-line cleanouts is a function of the design of the system. For example, a pure dendriform design would require only one terminal cleanout arrangement, while a multiple-cluster feeder design would require a terminal cleanout for each cluster. Cleanouts and/or shutoff valves should be provided at all major pipe junctions and at locations where pipe sizes change. The need for further cleanouts and/or shutoff valves may be related to the available cleaning methods, contingencies required for the system, and the projected use or growth rate of the service population.

The previous discussion of in-line cleanouts is pertinent to shutoff valves and bypassing arrangements. Pressure sewer main segment isolation for repair is necessary and the longer distance between valve stations makes isolation more difficult. However, the use of an arbitrary rule of maximum separation distance is often unnecessarily restrictive. The State of Florida[13] has therefore recommended that, in the absence of special needs for mechanical cleaning, pipe size changes, pipe bends in direction, or major main confluences, the spacing of inline shutoff valves need not be less than every 600 feet in high density areas and not more than 1000 feet in low-density areas.

Although a properly designed effluent pressure sewer system may never require flushing, the system layout should include end-of-line cleanouts and possibly in-line cleanouts. This will allow easy entry for flushing by some mechanical or pressurized means. Cooper Communities,

Inc. has established a policy whereby cleanouts are located at significant low points within the piping system.[65] This policy was established due to the thought that if solids buildup did occur, it would most likely occur at these low points.

Grinder pressure sewer systems will require periodic flushing when the daily flow from the homes is not sufficient to maintain scouring velocities. This would normally be a problem in a new development when just a few homes are built and are discharging into the system. It could also present a problem in a seasonal development where very little flow occurs during certain portions of the year.

Both manual and automatic air release valves are currently being used. Automatic air release valves tend to require more maintenance when the system is operating below design flow. Odors may become a problem at these points and can be vented to a soil bed when possible.

Carlson and Leiser[61] have shown that moist loam soils were found to have excellent possibilities in the efficient and inexpensive removal of undesirable odorous gases from air streams. The process seems to be particularly adaptable to lift stations and other small installation sites where odors are likely to occur. This method has successfully been used for odor abatement at lift stations within the STEP pressure sewer system at Kalispell, Idaho.[67]

It is recommended that system valves be made of bronze. Extensive corrosion has occurred within both iron and brass valves used for pressure sewer systems.[67] Cooper Communities, Inc. has utilized epoxy coated resilient-seat gate valves for a period of approximately four years; however, no evaluation has been made of the success of this type of valve.

7.3 ON-LOT FACILITIES

The largest capital, operation and maintenance expenses of pressure sewer systems are related to the on-lot pressurization unit. Factors which should be considered in the design of the pressurization units include[20]:

1. Type of pressurization unit.
2. Storage volume.
3. Single or multiple service.
4. Location of pressurization unit.
5. Service connection to collection system.
6. Electrical considerations.
7. Alarm and control system.
8. Aesthetics and safety problems.
9. Serviceability of components.
10. Materials of construction.
11. Contingency planning.

7.3.1 Type of Pressurization Unit

In the design of a pressure sewer system, the most important step is the selection of the type of pressurization unit. This will have specific effects on the remainder of the system. Unless circumstances prohibit one or the other of the two types, both grinder pump and STEP systems should be considered.

7.3.2 Pumps

On-lot pumping components are the heart of the pressure sewer system. These components must be thoroughly investigated and designed to function as a coherent system for the system to function effectively. STEP components normally include the interceptor tank, for removing as much grease and solids as possible; the pumping chamber, which houses the effluent pump; the effluent pump itself; discharge piping, including check and discharge shutoff valves; control systems for turning the pumps on and off;

high level alarms; overflow device; and service lines (Figure 3-2).

Effluent pumps should be made of cast iron, bronze, and/or plastic construction of the centrifugal type with an oil-filled submersible motor. Effluent pump ratings range from 1/4 to 2 horsepower, depending upon dynamic head and flow capacity requirements. Effluent pumps with ratings up to 3/4 horsepower can operate on 115 or 230 volt sources while effluent pumps with ratings over 3/4 horsepower require 230 volt service. Effluent pumps are capacitor start with either permanent split capacitor or split phase motors. Effluent pump motor starters and capacitors can be located in the motor or adjacent housing. Either a control box housing or a junction box is required to connect the pump and level controls to the service users power source.

For grinder pump systems, submersible centrifugal or semi-positive displacement (SPD) progressing cavity screw-type pumps having ratings within the range of 1 to 2 horsepower are normally specified. The performance for the two types of pumps are different in respect to capacities and shutoff head. The submersible centrifugal pump has generally a higher flow producing capacity at low heads while the SPD pump has the ability to generate higher pressures and more predictable flows at higher heads. For clustered service connections, submersible centrifugal grinder pumps are available with ratings between 3 and 7½ horsepower. The SPD pump is not presently available with ratings exceeding 1 horsepower.

Grinder pump units must be capable of comminuting all material normally found in domestic or commercial wastewater, including reasonable amounts of foreign objects such as glass, eggshells, sanitary napkins, and disposable diapers, into particles that will pass through the 1¼ inch standard discharge piping and downstream valves. Stationary and rotating cutter blades on bases should be made of hardened stainless steel.

EQUIPMENT AND MATERIAL 83

Single-phase motors are available in 208 or 230 volts and should be of the capacitor start/capacitor run type for high starting torque. Three-phase motors are available in 208, 230, 460, or 575 volts. All grinder pumps should be commercially shop-tested and should include visual inspection to confirm construction in accordance with the specifications for correct model, horsepower, cord length, impeller size, voltage, phase, and hertz. The pump and seal housing chambers should be tested for moisture and insulation defects. After connection of the discharge piping, the grinder pump should be submerged, and amperage readings taken in each electrical lead to check for an imbalanced stator winding.

Grinder pump components include: a holding basin with sufficient volume to accumulate enough liquid for a 1- to 2-minute grinder pump cycle operation and additional volume to hold emergency overflow; the grinder pump, with a combination grinding and macerating unit attached to the bottom of the pump; discharge piping including check and discharge shutoff valves; control circuits and components to operate the units; an alarm system to indicate to the homeowner that a high level is exceeded in the storage tank; and on-lot piping between the pump tank unit and the mainline in the street (Figure 3-1).

Both the STEP and grinder pump units must have power supplied to the units either through a control cabinet mounted nearby or through a conduit directly into the chamber unit with the controls inside. Care should be taken to make certain that all conduits are sealed to prevent water and gases from flowing through the conduit. Service connections from the house for both types of systems are equivalent.

7.3.3 Electrical Control Systems

Since the single phase submersible centrifugal grinder

pump has a capacitor start type motor, the capacitors and start relays must be located in a separate control panel enclosure. This control panel can be located either outside (NEMA 3 enclosure) or inside (NEMA 1 enclosure) the service location. Most STEP Pumps have starters and capacitors located inside the motor housing and therefore do not require a separate control panel containing these components. However, it is generally recommended that all installations have a separate control panel for ease of access and for maximum exposure of the alarm system. In the interest of safety, it is recommended that the control panel enclosure be placed within sight of the pump wetwell.

The control panel should include, but not be limited to, a magnetic starter with an ambient compensated bimetallic overload relay. The relay should have a test button for simulation of overload trip and a manual reset button. Fault protection should be provided with a molded case magnetic circuit breaker with internal common trip or multiple poles. A hand-off-automatic toggle switch for hand operation with a green light to indicate the pump-running mode should be provided for each pump and should be mounted on a bracket inside the control panel enclosure. As an alternative addition, Bowne[73] has recommended that a pump hour meter be considered for inclusion within the control panel in order to facilitate future monitoring of the pressure sewer system.

Should there be a power failure, pump malfunction, or flooded wetwell, pump controls and wiring must be accessible and must comply with all code regulations to insure safety of the service user or operating personnel. As an alternate, an explosion-proof combination motor control/junction box may be installed inside the wetwell.

Semi-positive displacement pumps having the starter and capacitor located in the pump core require only a standard junction box hook-up to the power source.

Most grinder pump applications require either a 208- or

EQUIPMENT AND MATERIAL 85

230-volt single phase power source, and the designer must be assured that this power requirement is compatible with the service user's power distribution system.

It is recommended that an audio and/or visual high water alarm be utilized with both the grinder pump and STEP systems. The purpose of this alarm is to alert the service user of a system malfunction and to alert him to call the service authority. The alarm should be designed so that the service user can reset the audio alarm after a malfunction, but not disable it for future malfunctions. The alarm system can be mounted outside or inside the service location. In some cases, one alarm will be installed inside the service location with a backup alarm located outside the service location.

Most pressurization units receive electrical power from the home being served. This is highly recommended because single house pressurization units seldom use enough electricity to meet the minimum billing charges of most power companies.

Electrical connections between the homeowner's panel, the pressurization unit panel, and the pressurization unit itself must be made according to local codes. Approved underground wiring is recommended for both pump and control circuits. As stated earlier, pump and control circuitry should have separate fuses within the control panel. The control panel should be located within site of the pressurization unit, should be lockable, and should be tamper proof. All electrical connections, particularly with the pressurization unit, must be watertight.

The following items should be included in the design of the control panel:

1. Pump and control devices should be on a separate circuit from the remainder of the house.
2. Inside the control panel, the pump circuit should have a separate fuse smaller than the homeowners breaker.

86 PRESSURE SEWER SYSTEMS

3. Electrical requirements should be compatible with the house system.
4. Audible alarm reset should be easily accessible and operative by the homeowner.

7.3.4 Single Or Multiple Service Pressurization Units

Economics tend to favor multiple service pressurization units where the served units are within a reasonable proximity of each other. However, the problems resulting from multiple service often outweigh the economic benefits. Some of the problems include:

1. Longer service lines with more opportunity for infiltration.
2. If the home supplying power becomes vacant, power will be shut off.
3. It is difficult to assess the proper credit to the home supplying power to the pressurization unit.
4. More reserve space is necessary.
5. If homes are at different elevations, the lower ones could be flooded if the pressurization unit malfunctions.
6. Assessing blame for damage to the system.

Cooper Communities, Inc. has found that the economics for multi-family dwellings outweigh the problems mentioned above and multiple service STEP pressurization units have subsequently been developed for up to six dwelling units.[64] In addition, duplex grinder stations have been employed that serve up to 99 dwelling units.[74]

7.3.5 Location Of The Pressurization Unit

The pressurization unit should normally be as close to the house as feasible. This reduces the length of house service and the associated infiltration. The control panel should be located on the side of the house closest to the pressurization unit, thereby reducing the cost of electrical connections and assuring quick access during an emergency. Consideration must also be given to pumping out the pressurization unit for STEP systems, and the pressurization unit should therefore be located in such a position that it can be pumped out.

7.3.6 Service Connection To The Collection System

The service lateral from the pressurization unit is generally a 1-inch to 1½-inch plastic pipe that connects the pressurization unit to the collection system. This service lateral is usually PVC, polyethylene, or polybutylene tubing with PVC pipe being the most widely used. Since the collection system is pressurized, a check valve is required to prevent backflow into the pressurization unit. In addition, gate or ball valves are used to allow isolation of the pressurization unit.

7.3.7 Aesthetics And Safety Problems

One of the most commonly heard complaints about pressure sewer systems is the aesthetics of the control panel and pressurization unit cover. When designing these items, the engineer should attempt to choose panels and covers that are as attractive or are as unobtrusive as possible. One very large safety concern is the pressurization unit cover. It is anticipated that this cover could be an attraction to children and it is recommended that a locking device be included. The control panel should, of course, be lockable and tamper proof.

7.3.8. Serviceability Of Components

Serviceability of the pressurization unit components is important to minimize down time of the unit and keep the cost of inspection and maintenance to a minimum. Quick disconnect features are recommended for both the piping and electrical connections to facilitate easy removal, inspection, and repair or replacement.

7.3.9 Materials Of Construction

The atmosphere in both STEP and grinder pump pressurization units can be very corrosive. It is incumbent upon the engineer to utilize materials that are capable of withstanding this atmosphere. Grinder pump systems are generally packaged in such a manner that these considerations have been incorporated at the factory. STEP systems, however, are often designed and fabricated locally; therefore, these considerations must be included in the design. All components of the STEP system exposed to the atmosphere (not always submerged) must be highly resistant to corrosion. Materials which have been acceptable for the different components are as follows:

1. Interceptor tank and pump basin—concrete, coated steel, fiberglass, and plastic.
2. Valves—bronze and plastic.
3. Ancillary items—plastic and stainless steel.
4. Pump housing—coated cast iron, bronze, and plastic.

5. Pump Impeller—plastic, bronze, and cast iron.
6. Tank and basin cover—concrete, fiberglass, and coated steel.

7.3.10 Other Provisions For On-Lot Pressurization Units

Farrell[72] has recommended the following alternatives for overflow provisions within pressurization unit systems: 1) connection to an abandoned septic tank, 2) construction of a special soil absorption system, 3) providing a sealed holding tank that can be pumped out after the overflow occurs, or 4) a sealed overflow pipe which causes plumbing fixtures to back up just as they would in a gravity system with plugged lines.

A holding tank may be installed adjacent to the pressurization unit and connected as an overflow device. The addition of a holding tank will reduce the cost-effectiveness of the entire system and will require the holding tank contents to be removed when the tank has filled. The contents of the holding tank will have to be removed by pumping into a tank truck or returned into the inlet of the pressurization unit.

If an existing on-site septic tank is available and is in good condition, it may be emptied, inspected, and rehabilitated as necessary for use as a holding tank. This is based on the assumption that the grinder pump system service can be restored within a reasonable time period. Emergency overflow capacity in a grinder pump wetwell should contain 6 to 8 hours of storage depending upon usage. Emergency overflow in STEP interceptor tanks should contain 24 hours of flow, and this usually is a function of the design of the interceptor tank. An advantage of the STEP system concept over the grinder pump system is the additional storage capacity available. As stated previously, the overflow of the STEP interceptor tank should equate to about 24 hours of available storage. Since most of the settleable and floatable solids have remained in the septic tank, the clarified effluent can be discharged into emergency overflow drainfields, similar to soil absorption systems utilized with septic tanks.

If a previously existing drainfield is in reasonable condition, the overflow from a STEP system may be connected to the drainfield for emergency usage. It should be noted that this can present a potential problem if the seasonal water table or flooding conditions in the area were the cause of the original drainfield failure. In these instances, ground or surface water could backflow into the pump wetwell thus generating infiltration into the entire system.

If a previously existing drainfield is unacceptable or unavailable, a new drainfield can be provided if required but it will be susceptible to the same conditions noted for existing drainfields. It is also possible that drainfields can assist in the destruction of odors within pressure sewer systems. Studies have shown that soil beds do effectively control sewage odors.[61]

Cooper Communities, Inc.[27] has successfully used a combination soil bed vent field and overflow drainfield on their STEP pressure sewer systems. However, it should be stressed that each installation should be checked for potential groundwater infiltration if such a vent field is utilized. In addition, it is recommended that a study be performed on the soils to make certain that they can treat odors and that they will supply sufficient treatment during the rare situation when overflow does occur.

It is not recommended to discharge grinder pump effluent into soil absorption systems. The solids and grease contained within grinder pump effluent will quickly clog soil pores and will subsequently create a situation where the soil absorption system is no longer usable.

8
Characteristics of Pressure Sewage

The guidelines from the State of Florida give the following pressure sewer characteristics as shown in Table 8-1[13]:

Table 8-1. Treatment Facility Influent Characteristics

Parameter	GP Systems		STEP Systems	
	Average	Range	Average	Range
BOD mg/l	350	300–400	143	110–170
TSS, mg/l	350	300–400	75	50–100
FLOWS, gpcd	70		70	

Further review of available literature for STEP Systems by Bowne[11] has revealed an EPA report by Kreissl[53] which quoted flow quantities at 64 gallons per capita per day (gpcd) and 50 and 70 percent reductions for BOD and Total Suspended Solids (TSS) respectively following septic tank treatment. Weibel, Straub, and Thoman[54] in their study of household sewage disposal systems reported BOD and TSS reductions on the order of 65 percent following septic tank treatment. Removals of 54 and 80 percent[55] were found by researchers at the University of Wisconsin, with flows of 32 gpcd.[56] Farrell et al.[57] reported flows ranging from 24 gpcd to 78 gpcd in their study of grinder pump pressure sewers. The Commission on Rural Water[58]

92 PRESSURE SEWER SYSTEMS

recommended the rate of 58 gpcd, and Bowne[11] also reported that Schmidt of General Development Corporation in Florida measured flows from a pressure sewer system at 40 gpcd, with BOD reductions of 78 percent.

Bowne[11] also tabulated the following after his comprehensive study of STEP system effluents:

Table 8-2. Selected Characteristics of Pressure Sewers

Flow	60 gpcd
BOD	0.07 ppcd
Total Suspended Solids (TSS)	0.07 ppcd

gpcd-gallons per capita per day
ppcd-pounds per capita per day

Bailey et al.[60] tabulated the following after experiments on septic tank influents and effluents at the Sanitary Engineering Center of the University of California:

Table 8-3. Average Septic Tank Influent and Effluent Characteristics

Parameter	Raw Sewage	Septic Tank Effluent	% Removal
BOD	150	75	60
COD	310	160	48.4
TSS	185	50	73
Volatile Solids	265	160	39.6

In addition, effluent coliform count may be higher than 10 per milliliter.

Bowne[77] has tabulated the following from a pressure sewer system containing 650 STEP units and 12 grinder pump units:

Table 8-4. Wastewater Characteristics of The Glide, Oregon Pressure Sewer System

Constituent	Effluent Samples	Mean	Standard Deviation	Grinder Samples	Mean	Standard Deviation
DO	161	0.3mg/l	0.2	32	0.5mg/l	0.3
TSS	139	52mg/l	20	44	226mg/l	86
BOD	120	118mg/l	49	46	304mg/l	87
Grease	14	16mg/l	15	6	42mg/l	16

The same literature provides the following measured wastewater characteristics:

Table 8-5. Selected Wastewater Characteristics for STEP and Grinder Pump (GP) Systems

Parameter	STEP Wastewater	GP Wastewater
Flow	48 gpcd	48 gpcd
BOD	0.047 ppcd	0.122 ppcd
TSS	0.021 ppcd	0.090 ppcd

As shown above, an average rate of flow for this same system was determined to be 48 gallons per capita per day. This was based upon a measured average flow rate of 150 gallons per day per dwelling unit with an average of 3.1 persons per dwelling unit. Bowne[76] has subsequently stated that the mean TSS and BOD values within the STEP system have been reducing with time.

At the grinder pump pressure sewer system in Albany, New York, hydrogen sulfide odors were detected, and concentrations of up to 2.5 mg/l were measured at the discharge of the pressure main.[5] In addition, the following characteristics were reported:

Table 8-6. Albany, New York Grinder Pump Wastewater Characterization

Parameter	Mean	Range
BOD	330	216–504
COD	855	570–1,450
TSS	310	138–468
Total Nitrogen	80	41–144
Total Phosphorus	15.9	7.2–49.3
Grease	81	31–140
pH units	—	7.1–8.7

Kreissl[4] tabulates the following which characterizes grinder pump effluent and compares it to normal household wastes and typical municipal wastes:

Table 8.7. Household Wastewater Characterization
(Units are in mg/l)

Parameter	Without Grinder	With Grinder	Typical Municipal Waste
BOD	415	465	200
TSS	296	394	200
Volatile SS	222	309	150
Total Nitrogen	51	52	40
Ammonia	11	10	25
Total Phosphorus	33	32	10
Grease	123	129	100

9
Operation and Maintenance

Kreissl[4] has stated that pressure sewers should only be considered with properly conceived management arrangements. Failure to do so could seriously limit the effectiveness of the pressure sewer system. It is therefore recommended that the design of any pressure sewer system include provisions for an operating authority as well as a plan for the operation and maintenance of the system.

It is generally recommended that pressure sewer systems be owned, operated, and maintained by a sewer district that would have the authority to assess monthly sewer charges, and the capability to provide service and maintenance. Municipalities and other government agencies are obviously acceptable to own, operate, and maintain pressure sewer systems, and property owner's associations have also proven acceptable when they have been given sufficient prior contractual authority. Both Arkansas and Tennessee have accepted property owner's associations as the operating authorities for pressure sewer systems.[26,27] Private corporations have also proven acceptable when proper contractual authority has been established. Arkansas[65] and Florida[21] have both allowed private corporations to own, operate, and maintain pressure sewer systems.

The most important consideration in installing an effluent pressure sewer system, next to its cost effectiveness, is the establishment of an effective long-term operation and maintenance program. Once a pressure sewer system is installed, the homeowner must rely on the pump to discharge the wastewater generated from the home. Consequently, the pump must be reliable and, when a failure does occur, it must be put back into operation in a reasonable period of time.

Operation and maintenance of a pressure sewer system is two faceted: (1) operation and maintenance for each onsite pressurization unit; and (2) operation and maintenance of the pressure mains. Simplicity of design limits the need for maintenance by reducing the number of possible system malfunctions. The utility company should provide a capable staff with a good service call system to guarantee prompt service. Routine maintenance inspection is desirable as a preventive measure.

Overton[20] has recommended that routine inspection of the pressurization unit should include the following:

1. Remove the pump from the chamber, check the pump intake for blockage, check the suction plate and body for corrosion, and clean the entire pump.
2. Check all valves for proper operation.
3. Return pump to chamber and test.
4. Test alarm system.
5. Check sludge depth for need of removal.

In addition, Overton[20] recommends that routine maintenance of the collection system is also desirable as a preventive measure. These inspections should include:

1. Exercising all valves on a periodic basis.
2. Flushing air release assemblies.
3. Flushing of mains, particularly at the end of lines.

It has been stated that pressure sewer pumps should last from ten to fifteen years before major overhaul is required. In any given year, there will be some pump failures, but the percentage would be very low. Experience has shown that more pressure sewer service calls result from failures and malfunctions of the other pumping system components such as the valves, level controls, etc. These components are just as important as the pump. If they don't function properly, the pump will not perform as expected, and may cause system failures. Therefore, it is important that the components be functional and of good quality.

It is recommended that a quantity of spare pumps and system components along with a parts inventory be kept on hand for service backup. The normal practice would be to replace a pump that has failed with a new pump or rebuilt pump. The pump that has failed would be taken to the service maintenance building for repair, and placed in stock for the next service call. The recommended number of spare pumps to have on hand can be equated to $3\%-5\%$ of the total pumps in service. Small systems should require a larger percentage of spare pumps and parts than will larger systems. In addition, more spare pumps would be required in the ten- to fifteen-year period of use in comparison to the first few years of service.

It is also recommended that a comprehensive public education program be developed and that this program be utilized and refined to the greatest degree possible. Experience has shown that an educated public will help to better maintain any proposed type of pressure sewer system. Public education combined with a comprehensive operation and maintenance plan should be an integral part of the design of any pressure sewer system.

10
Existing Design Manuals and Publications

At the current time, only one regulatory agency has published a definitive set of guidelines for the design of pressure sewer systems. These guidelines were published by the State of Florida Department of Environmental Regulation, and they are entitled "Design and Specification Guidelines for Low Pressure Sewer Systems."[13] These guidelines were written by a technical advisory committee which consisted of numerous experts in the field of pressure sewer design. These guidelines are comprehensive and have been widely accepted with a few minor exceptions.

Bowne, a member of the technical advisory committee which prepared the aforementioned guidelines for the State of Florida, has written a memorandum[14] outlining comments and exceptions he has for those guidelines. These comments and suggestions should be widely considered, especially considering Mr. Bowne's experience and expertise in the field of pressure sewers.

The Water Pollution Control Federation has recently published a manual which addresses pressure sewer systems in detail. It is entitled "Alternative Sewer Systems," and it has been designated as MOP (Manual of Practice) FD-12.

Kreissl has written an excellent review of pressure sewer design procedures in *Status of Pressure Sewer Technology*[4] as published by the U.S. Environmental Protection Agency. In this report, numerous references are quoted for design flows, pipe size selection, and pump selection.

The only other publication currently available from a regulatory agency is written by David M. Cochran[15] of the Texas State Department of Health. This document does not go into recommended design procedure, however, it does delve into various design considerations, especially as they relate to regulatory agencies. This report also provides brief operating experiences at three pressure sewer installations in Texas.

Each of the major pump companies involved with the manufacturing of pressure sewer pumps has prepared or has helped in the preparation of design manuals. Design manuals of note have been prepared by Peabody Barnes, Inc.,[2] Hydr-O-Matic Pump Company,[16] Environment One Corporation,[17] and F. E. Myers Company.[18] The design manual produced for the Hydr-O-Matic Pump Company was written at the Battelle Institute in Columbus, Ohio, and this manual has been utilized as a basis for much of the subsequent design methodology. In addition, the F. E. Myers Company has developed a computer program to assist in the design of pressure sewer systems.[19]

As stated earlier, Bowne has consistently been one of the most quoted authors concerning the design of pressure sewer systems. Bowne's publications continue to serve as the basis of many pressure sewer designs.

Robert Langford has written a report entitled *Effluent Pressure Sewer Systems*[9] which provides many excellent considerations for the design of pressure sewer systems.

Johnnie Overton has prepared a report on pressure sewer systems which touches upon his preferred design procedure.[20] In addition to Mr. Overton's report, ITT Community Development Corporation has also prepared an en-

EXISTING DESIGN MANUALS 103

gineering reort on the proposed septic tank effluent pumping system for their Palm Coast Development in Florida.[21] In that engineering report, the STEP system was actually called a Pretreatment Effluent Pumping (PEP) system in obvious reference to the treatment provided by the interceptor or septic tank.

Bibliography

1. "Pressurized Alternative Wastewater Systems," F. E. Myers Company, Ashland, Ohio, Newsletter Number 4-84.
2. Brinley, Robert K., Olmstead, R. D., and Wilkinson, Steven M., *Design Manual for Pressure Sewer Systems,* Peabody Barnes, Inc., Mansfield, Ohio.
3. Cliff, M. A., "Experience with Pressure Sewerage," Journal Sanitary Engineering Division—ASCE, 94, No. 5, pp. 849–865, 1968.
4. Kreissl, James F., *Status of Pressure Sewer Technology,* Report prepared for the U.S. Environmental Protection Agency Technology Transfer, Design Seminar for Small Flows, EPA Report Number 625/4-77-011, Volume 7, October, 1977.
5. Carcich, I. G., Hetling, L. J., and Farrell, R. P., *A Pressure Sewer Demonstration,* U.S. Environmental Protection Agency Report Number R2-72-091, 1972.
6. Mekosh, G., and Ramos, D., *Pressure Sewer Demonstration at the Borough of Phoenixville, Pennsylvania,* U.S. Environmental Protection Agency Report Number R2-73-270, 1973.
7. Hendricks, G. F., and Rees, S. M., *Economical Residential Pressure Sewage System with No Effluent,* U.S. Environmental Protection Agency Environmental Protection Technology Series Report Number EPA-600/2-75-072, December, 1975.

8. Rose, C. W., "Rural Wastes: Ideas Needed," *Water and Wastes Engineering,* 9, Number 2, pp. 46–47, 1972.
9. Langford, Robert E., *Effluent Pressure Sewer Systems,* Proceedings of the Water Pollution Control Federation Annual Conference, October, 1977.
10. Bowne, W. C., Consulting Engineer, *Pressure Sewer Systems,* 2755 Warren, Eugene, Oregon, May, 1974.
11. Bowne, W. C., Consulting Engineer, *Glide-Idleyld Park Sewerage Study,* 2755, Eugene, Oregon, 1975.
12. Winzler and Kelley Consulting Engineers, *Design—Construction Report for the Septic Tank Effluent Pumping Sewerage System,* Manila Community Services District, A report to the California State Water Resources Control Board, November, 1979.
13. *Design and Specification Guidelines for Low Pressure Sewer Systems,* Prepared by a Technical Advisory Committee for the State of Florida Department of Environmental Regulation, June, 1981.
14. Bowne, W. C., Consulting Engineer, "Memorandum Referencing *Design and Specification Guidelines for Low Pressure Sewer Systems,* State of Florida Department of Environmental Regulation, 1981," 2755 Warren, Eugene, Oregon 97405, July 3, 1981.
15. Cochran, David M., *Pressurized Sewer Systems-Regulatory Agency's Viewpoint,* Proceedings of the Water Pollution Control Federation Annual Conference, October, 1974.
16. Flanigan, L. J., and Cudnik, R. A., *Review and Considerations for the Design of Pressure Sewer Systems Utilizing Grinder Pumps,* Battelle Institute, Columbus, Ohio, Published by Hydr-O-Matic Pump Company, Ashland, Ohio, 1974.
17. *Design Handbook for Low Pressure Sewer Systems,* Third Edition, Environment One Corporation, Schenectady, New York, 1973.
18. *Myers Engineering and Design Manual,* F. E. Myers Company, Ashland, Ohio.

19. *Computer Programs and Engineering Criteria for Designing the Pressure Sewer System Using Grinder or Effluent Pumps,* F. E. Myers Company, Ashland, Ohio.
20. Overton, Johnnie M., *Introduction and Design of Pressure Sewer Systems,* ITT Community Development Corporation, Prepared for the University of Wisconsin Extension, March, 1981.
21. *Engineering Report of Pretreatment Effluent Pumping (PEP) Sewage System at Palm Coast, Florida,* ITT Community Development Corporation, Palm Coast, Florida, July, 1978.
22. Blaylock, Threet, and Associates, Inc., Consulting Engineers and Land Surveyors, *Revised Master Plan for Sewerage Facilities, 1977, Hot Springs Village, Arkansas,* Little Rock, Arkansas, March, 1977.
23. Bowne, W. C., *Report on Final Design Conference— Pumped Effluent Sewer System-Hot Springs Village,* Prepared for Cooper Communities, Inc., March, 1977.
24. Cooper Consultants, A Division of Cooper Communities, Inc., Consulting Engineers, Planners, and Surveyors, *Engineering Report—Escocia Subdivision Pumped Effluent Design-Hot Springs Village, Arkansas,* Prepared for Cooper Communities, Inc., Bella Vista, Arkansas, 1977.
25. Cooper Consultants, A Division of Cooper Communities, Inc. Consulting Engineers, Planners, and Surveyors, *Escocia Monitoring Systems—Escocia Subdivision-Hot Springs Village, Arkansas,* Prepared for Cooper Communities, Inc., Bella Vista, Arkansas, September, 1982.
26. Garver and Garver, Inc., Consulting Engineers and Planners, *Revised Master Plan for Water and Sewer, 1983, Hot Springs Village Arkansas,* Report to the Arkansas Department of Health, March, 1983.
27. Cooper Consultants, A Division of Cooper Communities, Inc., Consulting Engineers, Planners, and Surveyors, *A Master Plan for Wastewater Collection*

and Transmission at Tellico Village, Tennessee, Prepared for the Tellico Village Property Owners' Association, Bella Vista, Arkansas, August, 1984.
28. Crafton and Tull Consulting Engineers, Inc., *Master Plan for the Southeast and West Areas of Bella Vista Village,* Report to the Arkansas Department of Health, March, 1972.
29. Blaylock, Threet, and Associates, *Master Plan-Water Facilities, Cherokee Village, Arkansas,* Report to the Arkansas Department of Health, February, 1970.
30. Siegrist, R. L., "Minimum Flow Plumbing Fixtures," *Journal of the Environmental Engineer's Division,* Proceedings of ASCE, pp. 342–347, June, 1976.
31. Siegrist, R. L., Witt, M., and Boyle, W. C., "Characteristics of Rural Household Wastewater," *Journal of the Environmental Engineer's Division,* Proceedings of ASCE, pp. 533–548, June, 1976.
32. Maddaus, W. O., Parker, D. S., and Hunt, A. J., "Reducing Water Demand and Wastewater Flow," *Journal AWWA,* pp. 330–335, July, 1983.
33. Cole, C. A., and Sharp, W. E., "Impact of Water Conservation on Residential Septic Tank Effluent Quality," *On-Site Sewage Treatment,* Proceedings at the Third National Symposium on Individual Small Community Sewage Treatment, Chicago, Illinois, pp. 139–149, December, 1981.
34. Palmini, D. V., and Shelton, T. B., "Noncrisis Use of Household Water-Saving Devices," *Journal AWWA,* pp. 336–341, July, 1983.
35. Maddaus, W. O., and Feuerstein, D. L., "Effects of Water Conservation on Water Demands," *Journal of the Water Resources Planning and Management Division,* Proceedings of the American Society of Civil Engineers, New York, pp. 343–351, September, 1979.
36. Milne, M., *Residential Water Conservation,* University of California at Los Angeles, Report Number 35, California Water Resources Center Project UCAL-WRC-W-424, March, 1976.

37. Unibell Plastic Pipe Association, Dallas, Texas, *Handbook of PVC Pipe,* pp. 188–191, October, 1977.
38. Gray, Glenn C., "Environmental Constraints Challenge Designers of Shoreline Community Near Kansas City, Missouri," *Professional Engineer,* pp. 42–44, June, 1975.
39. Kent, J. C., *The Entrainment of Air by Water Flowing in Circular Conduits with Drowngrade Slopes,* Thesis, University of California, 1952.
40. Burton, L. H., and Nelson, D. F., *Surge and Air Entrainment in Pipelines,* Paper presented at conference: Control of Flow in Closed Conduits, Colorado State University, 1970.
41. Whitsett, A. M., *Practical Solutions to Air Entrainment Problems,* Paper presented at conference: Control of Flow in Closed Conduits, Colorado State University, 1970.
42. Winn, W. P., *Hydraulic Problems Associated with System Design and Operational Control of Flow in Large Transmission Systems,* Paper presented at conference: Control of Flow in Closed Conduits, Colorado State University, 1970.
43. American Society of Civil Engineers, "Combined Sewer Separation Publication Using Pressure Sewers," FWPCA Publication Number ORD-4, October 1969.
44. Flanigan, L. J., and Cadmik, C. A., "Pressure Sewer System Design," *Water and Sewage Works,* pp. R25–R34, and R87, April, 1979.
45. Neale, L. C., and Price, R. E., "Flow Characteristics of PVC Sewer Pipe," *Journal of the Sanitary Engineering Division,* Proceedings of the ASCE, 90, 109, 1964.
46. AWWA Standards Committee on Plastic Pipe, "Plastic Pipe and the Water Utility," *American Water Works Association Journal,* 63, 352, 1971.
47. Lescovich, J. E., "Locating and Sizing Air Release Valves," *American Water Works Association Journal,* 64, 457, July, 1972.

48. Great Lakes-Upper Mississippi River Board of Sanitary Engineers, "Recommended Standards for Sewage Works," 1978 Revised Edition, Health Education Service, P.O. Box 7126, Albany, New York.
49. McPherson, M. B., Tucker, L. S., and Hobbs, M. F., "Minimum Transport Velocity for Pressurized Sanitary Sewers," ASCE Combined Sewer Separation Project, Technical Memorandum Number 7, November 1967.
50. Fair, G. M., Geyer, J. C., and Okun, D. A., *Elements of Water Supply and Wastewater Disposal*, Second Edition, John Wiley and Sons, Inc., New York, 1971.
51. Geyer, J. C., and Lentz, J. J. "An Evaluation of the Problems of Sanitary Sewer System Design," *Journal of the Water Pollution Control Federation,* 38 (7), 1138–1147, July, 1966.
52. Hendricks, G. F., "Pressure Sewage System and Treatment. Grandview Lake, Indiana," Demonstration Project Number S-801041, Ohio Home Sewage Disposal Conference, Ohio State University, Columbus, Ohio, January 29–31, 1973.
53. Kreissl, J. K., "Waste Treatment for Small Flows," Environmental Protection Agency, Cincinnati, 1971.
54. Weibel, S. R., Straub, C. P., and Thoman, J. K., "Studies on Household Sewage Disposal Systems," Part I, Cincinnati, Federal Security Agency, 1949.
55. Ziebell, W. A., Nero, D. H., Deininger, J. F., and McCoy, E., "Use of Bacteria in Assessing Waste Treatment and Soil Disposal Systems," paper presented at the National Symposium on Home Sewage Disposal by researchers of the Small Scale Waste Management Project, University of Wisconsin, 1975.
56. Witt, M., Siegrist, R., and Boyle, W. C., "Rural Household Wastewater Characterization," paper presented at the National Symposium on Home Sewage Disposal by researchers of the Small Scale Waste Management Project, University of Wisconsin, 1975.

57. Farrell, R. P., Watson, K. S., and Anderson, J. A., "The Contribution from the Individual Home to the Sewer System," *Journal of the Water Pollution Control Federation,* Volume 39, Number 12, 1967.
58. Goldstein, S. N., and Moberg, W. J., "Wastewater Treatment Systems for Rural Communities," Commission on Rural Water, Washington, D.C., 1973.
59. Weibel, S. R., Bendixen, T. W., and Coulter, J. B., *Studies on Household Sewage Disposal Systems,* Part III, Washington, D.C., U.S. Government Printing Office, 1955.
60. Bailey, J. R., Benoit, R. J., Dodson, J. L., Robb, J. M., and Wallman, H., *A Study of Flow Reduction and Treatment of Waste Water from Households,* Report Prepared for the Federal Water Quality Administration, Department of the Interior, Advanced Waste Treatment Research Laboratory, Cincinnati, Ohio, December, 1969.
61. Carlson, D. A., and Leiser, C. P., "Soil Beds for the Control of Sewage Odors," Journal WPCF, pp. 829–840, May, 1966.
62. Bowne, W. C., "Collection Alternative: The Pressure Sewer," Proceedings of the Third National Conference for Individual Onsite Wastewater Systems, National Sanitation Foundation, pages 171–186, 1977.
63. Cooper, I. A., and Rezek, J. W., *Septage Treatment and Disposal,* Report for the Environmental Protection Agency Technology Transfer Seminar Program on Small Wastewater Treatment Systems, 1977.
64. Cooper Consultants, A Division of Cooper Communities, Inc. Consulting Engineers, Planners, and Surveyors, "Water and Pumped Effluent Systems for Coronado Courts, Blocks 4, 5, and 6 in Hot Springs Village, Arkansas," Report to the Arkansas Department of Health, Bella Vista, Arkansas, March, 1978.
65. Cooper Consultants, A Division of Cooper Communities, Inc. Consulting Engineers, Planners, and

Surveyors, "Water and Pumped Effluent Design for Melanie Courts, Block 1, Including Partial Off-Site Pressure Sewer Design for the Metfield High Density Development Area, Bella Vista Village, Arkansas," Report to the Arkansas Department of Health, September 1980.

66. Gross, M. A., and Thrasher, D., *Causes, Correction, and Prevention of Septic Tank-Soil Absorption System Malfunctions,* Proceedings of the Fourth National Symposium on Individual and Small Community Sewage Systems, American Society of Agricultural Engineers, December, 1984.

67. Cooper, I. A., and Rezek, J. W., *Experiences with Pressure Sewer Systems,* Proceedings of the 51st Annual Water Pollution Control Federation Conference, Anaheim, California, October, 1978.

68. Bowne, W. C., *Variable Grade Gravity Sewers at Glide, Oregon* W. C. Bowne Consulting Engineer, 2755 Warren Eugene, Oregon, November, 1984.

69. Cooper Consultants, A Division of Cooper Communities, Inc., Consulting Engineers, Planners, and Surveyors, "Pumped Effluent Pressure Sewer Systems," Report to the Arkansas Department of Health, Bella Vista, Arkansas, June, 1976.

70. Tollefson, D. J., and Kelly, R. F., "STEP Pressure Sewers Are a Viable Wastewater Collection Alternative," *Journal of the Water Pollution Control Federation,* 53 (7), pages 1004–1014, July, 1983.

71. Sanson, Richard L., "Design Procedure for a Rural Pressure Sewer System," *Public Works,* 104 (10): pages 86–87, October, 1973.

72. Farrell, R. Paul, "Pressure Sewers and the Grinder Pump Which Makes Them Possible," *Journal of the New England Water Pollution Control Association,* November, 1972, pages 200–212.

73. Bowne, W. C., *Design Flows and Hydraulics of Centrifugal Pump Pressure Sewer Systems,* W. C. Bowne

Consulting Engineer, 2755 Warren, Eugene, Oregon, November, 1977.
74. Cooper Consultants, A Division of Cooper Communities, Inc., Consulting Engineers, Planners, and Surveyors, "Water and Sewer Extensions for Brompton Courts, Bella Vista, Arkansas," Report to the Arkansas Department of Health, Bella Vista, Arkansas, January, 1978.
75. Bowne, W. C., *Sewer Gas Considerations for Pressure Sewer On-Lot Facilities*, W. C. Bowne Consulting Engineer, 2755 Warren, Eugene, Oregon, May, 1978.
76. Bowne, W. C., *Septage Accumulation Rates, The Glide Pressure Sewer System*, W. C. Bowne Consulting Engineer, 2755 Warren, Eugene, Oregon, February, 1982.
77. Bowne, W. C., *Wastewater Characteristics of the Glide, Oregon Pressure Sewer System*, W. C. Bowne Consulting Engineer, 2755 Warren, Eugene, Oregon, February, 1982.
78. Bowne, W. C., *Two Phase Flow in Pressure Sewers and Small Diameter Gravity Sewers*, W. C. Bowne Consulting Engineer, 2755 Warren, Eugene, Oregon, April, 1983.
79. Rezek, Joseph W., and Cooper, Ivan A., *Investigations of Existing Pressure Sewer Systems*, Environmental Protection Agency Report Number EPA/600/2-85/051, April, 1985.

Glossary

Absorption Factor—The ultimate percentage of residences as compared to the total available homesites in a particular development.

Anaerobic Treatment—Stabilization of wastewater accomplished by the action of microorganisms in the absence of air or elemental oxygen.

BOD-Biochemical Oxygen Demand—A standard test used in assessing the strength of wastewater. It measures the strength of oxygen used in the biochemical oxidation of organic matter in a specified time, at a specified temperature, and under specified conditions.

Capital Replacement Cost—The cost of replacement for any fixed asset. In pressure sewer systems, this is commonly thought of as the replacement cost for pumps which have a shorter economic life than that of the pressure sewer system itself.

Centrifugal Pump—A pump consisting of an impeller fixed on a rotating shaft enclosed in a housing which has an inlet and a discharge connection. The rotating impeller creates pressure in the liquid by the velocity derived from centrifugal force.

Cleansing Velocity (Scouring Velocity)—In wastewater, the minimum velocity necessary to dislodge stranded material from the perimeter of the pipeline by the motion of the fluid.

COD-Chemical Oxygen Demand—A measure of the amount of oxygen required for the chemical oxidation of organic material in wastewater.

Conventional Pump Station—In wastewater, a pump station required within a conventional gravity sewer system.

Depreciation Cost—The cost of the loss in service value which is not restored by current maintenance and which is due to all the factors causing the ultimate retirement of the property.

Dissolved Oxygen—The oxygen dissolved in water or wastewater, usually expressed in milligrams per liter.

Dosing Siphon—A siphon which will automatically discharge the liquid accumulating in a tank to a piping system which will subsequently transmit the liquid to a wastewater treatment system.

Drainfield—*See* Soil Absorption System.

Economic Life—The life of a system or unit beyond which that system or unit will not be economically viable. Also defined as the number of years over which the depreciation cost can be spread.

Floatable Solids—Those solids contained in wastewater which will rise to the surface. *See* Grease.

Flow Equalization—The process of storing wastewater for release to a sewer system or treatment plant at a controlled rate to provide a reasonably uniform flow.

Flushing—The removal of deposits of material which has lodged in pipes because of inadequate velocity of flow. Water or wastewater is discharged into the pipe at such rates that the larger flow and higher velocity are sufficient to remove the material.

Flushing Station—An access point in a pressure sewer system where the process of flushing can be initiated.

Grease—In wastewater, a group of substances including fats, waxes, free fatty acids, calcium and magnesium soaps, mineral oils, and other nonfatty materials. These are water-insoluble organic compounds which can be removed by natural floatation skimming.

GLOSSARY

Headloss—Energy loss in a liquid due to the resistance of flow within the conduit or pipe system.

Hydraulic Gradient (Hydraulic Grade Line)—A hydraulic profile of the pressure level within a pipe at all points along the pipe. For closed conduits under pressure, it is the slope of the line joining the elevations to which water would rise in pipes freely vented and under atmospheric pressure. For pressure sewer systems, it is an expression of the pressure that must be overcome in order to inject liquid into a pipe at any given location.

Hydraulic Radius—The ratio of area to wetted perimeter in a cross-section of a conduit.

Infiltration—The quantity of groundwater that leaks into a pipe through joints, porous walls, or breaks.

Inflow—The quantity of water, especially surface water, which enters a pipe from other sources than infiltration.

Interceptor Tank—In pressure sewers, the tank which accepts raw wastewater from its source and holds that wastewater for a significant period of time before discharging. The Interceptor Tank is equivalent to a Septic Tank because it allows treatment by means of settling or floatation combined with anaerobic digestion.

Intermediate Pump Station—In pressure sewers, a pump station which is required to pump wastewater towards the treatment plant. The term intermediate is used to differentiate this pump station from individual pump stations or pressurization units used at each source of wastewater.

Multiple Service Pressurization Unit—A pressurization unit which serves more than one source of wastewater flow.

Nitrogen—An essential nutrient that is often present in wastewater as ammonia, nitrate, nitrite, and organic nitrogen. The sum of the nitrogen available in these forms is expressed as milligrams per liter of total nitrogen.

Non-Gradient Gravity Flow—In pressure sewers, the downhill flow of wastewater at atmospheric pressure.

Onsite Wastewater Disposal—The process of treating and disposing of a wastewater at its point of origin. Usually used to describe septic tank systems or other systems used to treat and dispose of wastewater at or near the homesite.

Peak Factor—The ratio of the maximum or peak flow in a given period of time to the overall average flow.

pH—A measure of the hydrogen concentration in a solution. A low pH (below 7) indicates acidity whereas a high pH (above 7) indicates alkalinity.

Phosphorus—An essential chemical element and nutrient which occurs as orthophosphate, pyrophosphate, tripolyphosphate, and organic phosphate forms. Each of these forms and their sum, total phosphorus, are expressed as milligrams per liter elemental phosphorus.

Pressure Sustaining Device—A device or valve used in pressure sewer systems to maintain pressure in downslope flow situations so that air will not enter into the piping system.

Pressurization Unit—The on-site pump station within pressure sewer systems which accepts raw wastewater from the source and pumps the wastewater into the pressurized piping system.

Pump Curve—A graph which shows the interrelation of dynamic head and flow capacity of a pump. Most pump curves will also contain information on motor speed, brake horsepower, and efficiency.

Scouring Velocity—In wastewater, the minimum velocity necessary to dislodge stranded material from the perimeter of the pipeline by the motion of the fluid.

Semi-Positive Displacement Pump—A type of pump in which the water is induced to flow from the source through an inlet pipe and into the pump chamber by a displacement device. This device is similar to a screw-type impeller and it operates in a rubber boot.

Septage—The sludge produced in septic tanks and in interceptor tanks which are utilized within pressure sewer systems.

Septic Tank—An underground vessel for treating wastewater from a source by a combination of settling, floatation, and anaerobic digestion. The effluent from a septic tank is usually discharged into a soil absorption system and the septage is pumped out periodically and transported to a treatment facility for disposal.

Settleable Solids—The matter in wastewater which will not stay in suspension during a preselected settling period and therefore settles to the bottom during that period.

Shutoff Head—The head or pressure at which a pump will no longer discharge flow.

Small Diameter Gravity Sewer—A gravity sewer system which has septic tanks or interceptor tanks at each source of wastewater. The septic tanks remove much of the grease and solids from the wastewater thus allowing smaller pipe to be used since there is less likelihood of clogging. In addition, smaller concentrations of grease and solids allow the designer to virtually eliminate the strict manhole and alignment requirements of conventional gravity sewers.

Soil Absorption System—A wastewater disposal system which allows septic tank effluent or other treated wastewater to percolate through an acceptable soil and therefore receive proper final treatment.

System Headloss Curve—A graph of the interrelationship of head (pressure) and flow for a given pipeline.

Total Suspended Solids (TSS)—Insoluble solids that are either in suspension within or float on the surface of water or wastewater. The reported quantity is removed from wastewater in a prescribed laboratory test and is referred to as nonfilterable residue.

Variable Speed Pump—A pump whose motor is capable of running at variable speeds which therefore allows the pump to discharge variable flow rates.

Vertical Turbine Pump—A centrifugal pump in which fixed guide vanes partially convert the velocity energy of the water into pressure head as the water leaves the impeller. Impellers on a vertical turbine pump can be stacked in such a way as to increase the head (pressure) of the discharge at any rate of flow.

Volatile Solids—Materials, generally organic, which can be driven off from a sample by heating, usually to 550 degrees Celsius.

Volatile Suspended Solids (VSS)—That portion of suspended solids, including organic matter and volatile inorganic salts, which will ignite and burn when heated to 550 degrees Celsius for 60 minutes.

Index

ASTM 77
Absorption Factor 40, 44, 48, 49
Activated Sludge Wastewater Treatment 8
Air Binding 51
Air Entrapment 25, 38, 57–60, 67
Air Release Valves 7, 25, 34, 50, 54, 58, 59, 80, 98
Air and Vacuum Release Valves 58, 59, 60, 62, 65, 66, 68, 73, 74
Alarm System 12, 13, 27, 31, 81–86, 98
Albany, New York 6, 7, 77, 94
American Society of Civil Engineers 5, 6, 105, 108–110
American Water Works Association 52, 108, 109
Ammonia 95
Anaerobic Digestion 30
Arkansas Department of Health 47, 53, 107, 108, 111–113
Average Flow 43, 44, 45, 48, 56, 91, 92, 94
BOD 20, 29, 30, 31, 91–95
Bar Screens 29
Battelle Institute 42, 43, 102, 106
Bella Vista Village, Arkansas 55, 107, 108, 111–113
Bend, Oregon 7
Bowne, Bill 9, 20, 29, 30, 32, 34, 41, 42, 47, 55, 57, 58, 73, 91, 92, 94, 101, 102, 106, 107, 111, 112, 113

Butt-Fusion Joints 79
COD 92, 94
California State Water Resources Control Board 9, 10, 106
Capital Cost 27, 37, 72, 73, 75, 80
Capital Replacement Cost 75
Centrifugal Pumps 6, 8, 21, 31, 34, 35–38, 39, 41, 42, 49, 52, 53, 61, 69, 70, 82, 83, 112
Cleanouts *See* Flushing Stations 79
Cleansing Velocity *See* Scouring Velocity 19, 39
Coeur d'Alene, Idaho 9
Columbia River 6
Columbus, Indiana 7
Comminuters 29
Commission on Rural Water 91, 111
Computer Design Methods 39
Construction Costs 18, 50
Controls 30, 31, 34, 81, 82, 83, 84, 85, 88, 99
Conventional Pump Stations 3, 17, 23, 24, 62, 63, 73
Coolin, Idaho 32
Cooper Communities, Inc. 8, 20, 42, 46, 47, 49, 55, 79, 80, 86, 90, 107, 111, 112, 113
Corrosion 34, 77, 80, 88, 98
Demographics 40, 46–49
Department of Health, Education, and Welfare 32
Depreciation Cost 75

122 PRESSURE SEWER SYSTEMS

Design Flow 39–52, 64, 67, 68, 69
Design Head 64, 68, 69
Dissolved Oxygen (DO) 93
Dosing Siphon 55
Douglas County, Oregon 9, 47
Drainfields *See* Soil Absorption Systems 32, 90
Durtschi, Kenneth 9
Dynamic Head 54
Easements 16, 18, 19, 64
Economic Life 75
Electricity Costs 17
Emergency Plan 22
Engineering Cost 16, 17
Environment One Corporation 35, 40, 41, 78, 102, 106
Environmental Protection Agency 6, 10, 102, 105, 110, 113
Explosion-Proof Requirements 71, 84
Extended Areation Wastewater Treatment 8
F. E. Meyers Pump Company 36, 54, 102, 105, 106, 107
Fair, Dr. Gordon M. 5
Farmers Home Administration 8
Federal Grants 2, 3
Federal Power Commission 35
Floatable Solids 90
Florida Department of Environmental Regulation 101, 106
Flow Equalization 37, 50
Flushing Stations *See* Cleanouts 24, 65, 66, 68, 74
General Development Corporation 8, 92
General Electric Company 6
Glide-Idleyld Park, Oregon 9, 32, 42, 47, 55, 93, 106, 112
Grandview Lake, Indiana 7, 8, 78, 110
Gravity Sewer Construction Cost 3, 4, 73, 75

Grease 7, 11, 20, 24, 30, 31, 38, 55, 56, 57, 81, 90, 93, 94, 95
Grit Chambers 29
Grit Removal 14
Harvard University 4
Hazen Williams "C" Factor 52, 67
Hazen Williams Headloss Formula 67
Headloss 25, 37, 44, 50, 53, 58, 60, 67, 68, 69, 72
High Density Polyethylene Pipe (HDPE) 78
Hot Springs Village, Arkansas 41, 42, 107, 111
Hydromatic Pump Company 36, 102, 106
Hydraulic Grade Line 54, 55
Hydraulic Gradient 64, 66, 67, 68, 69, 72
Hydraulic Radius 67
Hydrogen Sulfide 25, 28, 33, 94
ITT Community Development Corporation 102, 107
Infiltration 2, 7, 14, 16, 20, 22, 29, 34, 86, 87, 90
Inflation 76
Inflow 2
Interceptor Tanks 9, 11, 20, 25, 29, 30, 31, 33, 34, 35, 52, 81, 88, 89
Intermediate Pump Stations 21, 34, 70–72
Inventory 22, 99
Inventory–Pumps 99
Inventory–Spare Parts 99
Kalispell, Idaho 32, 80
Kreissl, James 20, 29, 30, 53, 57, 91, 94, 97, 102, 105, 110
Lakeshore Developments 15, 18, 19
Langford, Robert 15, 32, 41, 102, 106
Manholes 19, 23, 62, 63, 64, 65, 66, 68, 73
Manila, California 9, 106
Mercury Float Switches 31
Metfield Pressure Sewer System 55, 112

INDEX 123

Methane 33
Miami, Florida 8
Multiple Service Pressurization Units 27, 86
Myers Pump Company *See* F. E. Myers Pump Company 36
New York Department of Environmental Conservation 6
Non-Gradient Gravity Flow 60, 67
Odor 25, 28, 60, 80, 90, 94, 111
On-site Wastewater Disposal 2
Operating Authority 22, 23, 97
Operation and Maintenance Costs 14, 16, 24, 27, 28, 72, 75, 80
Operation and Maintenance Plan 22, 23, 97, 98, 99
Overton, Johnny 98, 102, 107
PVC Plastic Pipe 18, 52, 53, 77, 78, 87
Palm Coast, Florida 103
Peabody-Barnes Pump Company 36, 102, 105
Peak Factor 29, 43, 44, 46
Peak Flow 21, 22, 36, 37, 43, 44, 45, 46, 49, 50, 56, 64, 67, 72
pH 94
Phoenixville, Pennsylvania 7, 78, 105
Plastic Pipe *See* PVC Plastic Pipe 18, 52, 53, 77, 78, 87
Pocono Mountains, Pennsylvania 7
Polybutylene Pipe 78, 87
Polyethylene Pipe 87
Port Charlotte, Florida 8, 32
Port St. Lucie, Florida 8, 32
Portland, Oregon 6
Preliminary Settling 14
Pressure Sensing Switches 31
Pressure Sewer Construction Cost 72–76
Pressure Sustaining Devices 50, 54, 57
Pretreatment Effluent Pumping (PEP) System 103, 107

Priest Lake, Idaho 9
Primary Clarifiers 29
Probability Method 39, 40–42, 61, 69
Progressing Cavity Pump *See* Semi-Positive Displacement Pump 45
Public Education 22, 23, 99
Pump Curve 36, 37, 41, 54, 70
Pump Domination 49, 51
Pump Hour Meter 84
Pump Overheating 51
Pump Siphoning 59
Pump Stations *See* Conventional Pump Stations or Intermediate Pump Stations
Pump Vortexing 59
Radcliffe, Kentucky 5
Rational Method 39, 42–46, 61
Recommended Standards for Sewage Works 56, 58, 110
Recreational Communities 15, 46, 49
Replacement Cost 16, 51
Retirement Communities 15, 46, 49, 74
Rezek, Henry, Meisenheimer, and Gende, Inc. 10
Road Borings 19
Rock Excavation 18
Rose, Cecil 8, 106
Rubber Gasket Joints 77
Safety Factor 78
Schmidt, Harold 8, 92
Scouring Velocity *See* Cleansing Velocity 30, 36, 37, 55–56, 80
Semi-Positive Displacement Pumps 6, 9, 35–38, 39, 40, 41, 42, 49, 61, 69, 70, 82
Septage 25, 31, 32, 33, 113
Septic Tank 3, 8, 9, 11, 12, 13, 29, 30, 31, 32, 33, 89, 90, 91
Septic Tank Effluent 47, 56, 91, 92
Service Connections 19, 79, 81, 83, 87

Settleable Solids 20, 90
Shutoff Head 21, 36, 37, 51, 52, 68
Shutoff Valves 79, 81, 83
Sludge 31, 32
Small Diameter Gravity Sewer 55, 113
Soil Absorption System 3, 8, 10, 31, 89, 90
Soil Vent Field 34
Solids-Handling Submersible Pumps 6, 71, 72
Solvent Weld Joints 77, 78
Standpipes 57
Storage Volume 81, 83, 89, 90
System Headloss Curve 50, 54
System Maps 26
Tellico Village, Tennessee 21, 42, 49, 108
Tennessee Department of Health and Environment 47
Texas State Department of Health 102
Toran Pump Company 36
Total Coliform Count 92
Total Nitrogen 94, 95
Total Phosphorus 94, 95
Total Suspended Solids (TSS) 20, 29, 30, 31, 91, 92, 93, 94, 95
Trench Dewatering 18
Trench Shoring 19
USEPA 6, 10, 102, 105, 110, 113
U.S. Public Health Service 32
University of California Sanitary Engineering Center 92
Variable Speed Pumps 71
Ventilation 25, 30, 33, 34
Verticle Turbine Pumps 71
Volatile Solids 92
Volatile Suspended Solids 95
Wastewater Treatment 1, 2, 3, 7, 17, 20, 21, 22, 24, 29, 37, 50, 91
Water Conservation 40, 47, 48
Water Hammer 54, 78
Water Pollution Control Federation 101, 106, 110, 111, 112
Weatherby Lake, Missouri 78
Winzler and Kelly, Consulting Engineers 9, 106